Dragan Krecar

Physical Analytics in Materials Science

Dragan Krecar

Physical Analytics in Materials Science

Applications of High Performance Physical Analytics (SIMS, RBS, SEM/TEM and AES) in Materials Science

Südwestdeutscher Verlag für Hochschulschriften

Impressum/Imprint (nur für Deutschland/ only for Germany)
Bibliografische Information der Deutschen Nationalbibliothek: Die Deutsche Nationalbibliothek verzeichnet diese Publikation in der Deutschen Nationalbibliografie; detaillierte bibliografische Daten sind im Internet über http://dnb.d-nb.de abrufbar.

Alle in diesem Buch genannten Marken und Produktnamen unterliegen warenzeichen-, marken- oder patentrechtlichem Schutz bzw. sind Warenzeichen oder eingetragene Warenzeichen der jeweiligen Inhaber. Die Wiedergabe von Marken, Produktnamen, Gebrauchsnamen, Handelsnamen, Warenbezeichnungen u.s.w. in diesem Werk berechtigt auch ohne besondere Kennzeichnung nicht zu der Annahme, dass solche Namen im Sinne der Warenzeichen- und Markenschutzgesetzgebung als frei zu betrachten wären und daher von jedermann benutzt werden dürften.

Verlag: Südwestdeutscher Verlag für Hochschulschriften Aktiengesellschaft & Co. KG
Dudweiler Landstr. 99, 66123 Saarbrücken, Deutschland
Telefon +49 681 37 20 271-1, Telefax +49 681 37 20 271-0, Email: info@svh-verlag.de
Zugl.: Wien, Technische Universität Wien, Dissertation, 2006

Herstellung in Deutschland:
Schaltungsdienst Lange o.H.G., Zehrensdorfer Str. 11, D-12277 Berlin
Books on Demand GmbH, Gutenbergring 53, D-22848 Norderstedt
Reha GmbH, Dudweiler Landstr. 99, D- 66123 Saarbrücken
ISBN: 978-3-8381-0891-9

Imprint (only for USA, GB)
Bibliographic information published by the Deutsche Nationalbibliothek: The Deutsche Nationalbibliothek lists this publication in the Deutsche Nationalbibliografie; detailed bibliographic data are available in the Internet at http://dnb.d-nb.de.

Any brand names and product names mentioned in this book are subject to trademark, brand or patent protection and are trademarks or registered trademarks of their respective holders. The use of brand names, product names, common names, trade names, product descriptions etc. even without
a particular marking in this works is in no way to be construed to mean that such names may be regarded as unrestricted in respect of trademark and brand protection legislation and could thus be used by anyone.

Publisher:
Südwestdeutscher Verlag für Hochschulschriften Aktiengesellschaft & Co. KG
Dudweiler Landstr. 99, 66123 Saarbrücken, Germany
Phone +49 681 37 20 271-1, Fax +49 681 37 20 271-0, Email: info@svh-verlag.de

Copyright © 2008 Südwestdeutscher Verlag für Hochschulschriften Aktiengesellschaft & Co. KG and licensors
All rights reserved. Saarbrücken 2008

Produced in USA and UK by:
Lightning Source Inc., 1246 Heil Quaker Blvd., La Vergne, TN 37086, USA
Lightning Source UK Ltd., Chapter House, Pitfield, Kiln Farm, Milton Keynes, MK11 3LW, GB
BookSurge, 7290 B. Investment Drive, North Charleston, SC 29418, USA
ISBN: 978-3-8381-0891-9

No amount of experimentation can ever prove me right,
a single experiment can prove me wrong.

Albert Einstein
(1879-1955)

Danksagung

Mein besonderer Dank gilt meinem Betreuer Prof. Herbert Hutter erstens für die Möglichkeit der Durchführung sowohl der Diplom- als auch der Doktorarbeit und zweitens dafür, dass er auch wenn es um private und nichtwissenschaftliche Fragen ging, immer den besten Rat wusste.

Besonders möchte ich mich auch bei den Mitgliedern der Arbeitsgruppe Physikalische Analytik bedanken. An erster Stelle möchte ich mich hier bei Kurt Piplits und Dipl.-Ing. Karl Mayerhofer für unsere Erfolgreiche Zusammenarbeit bedanken. Weiters gilt mein Dank den Ehemaligen, Dr. Martin Rosner, Dipl.-Ing. Michael Fuchs und Dipl.-Ing. Jürgen Zwanziger, und den aktuellen Mitgliedern der Arbeitsgruppe: Dipl.-Ing. Johann Moser und Markus Hochegger für anregende Diskussionen über wissenschaftliche aber auch über unwissenschaftliche Fragen.

Ganz herzlich möchte ich mich bei Prof. Herbert Danninger und Dr. Vassilka Vassileva von der Arbeitsgruppe Pulvermetallurgie der TU Wien für die langjährige und erfolgreiche Zusammenarbeit bedanken.

Meinen weiteren Kooperationspartnern: Dipl.-Ing. Sven Peissl von der Montanuniversität Leoben, Prof. Peter Bauer von der Johannes Keppler Universität Linz und Dr. Reinhard Kögler vom Forschungszentrum Rossendorf möchte ich für die Zusammenarbeit und für das Bereitstellen der Proben, als auch für die anregenden wissenschaftlichen Diskussionen danken.

Dem österreichischen Fond zur Förderung der wissenschaftlichen Forschung FWF möchte ich für die finanzielle Unterstützung (FWF Projekte 14489 und 15931) während meiner Arbeit danken.

Bei allen meinen Studienkollegen des Jahrganges 1998, insbesondere Alina, Claudia, Mol, Jürgen, Martin, Bernhard, Robert, Michi und Matthias möchte ich mich für die wunderschönen 7 Jahre meines Lebens und natürlich für DAS Geschenk bedanken.

Meinen Freunden Tim, Daniel, Jan und Peter möchte ich auch danken, dass sie mir immer mit Rat und Tat zur Seite standen, nicht nur, wenn es um Literatur oder ähnliche Sachen ging.

Meiner Freundin Birgit danke ich für ihre Liebe, Motivation und die Unterstützung während den letzten Jahren. Bei ihren Eltern möchte ich mich auch für die moralische Unterstützung bedanken.

Den Familien Kunkic und Fajt gilt mein herzlicher Dank für Ihre Liebe und Ihre Unterstützung während meines Studiums.

Meinen Eltern, Dragica und Mihajlo, gilt mein unendlicher und herzlicher Dank, denn ohne Ihre Unterstützung während meines Studiums wäre diese Arbeit nicht möglich gewesen. Hvala za sve i ja vas volim!

Danke.

5	**RESULTS AND DISCUSSION**	**71**
5.1	POWDER METALLURGY	71
5.2	TRIBOLOGY	75
5.3	ION IMLANTATION AND DEFECT ENGINEERING	77
5.3.1	IMPURITY GETTERING – GETTERING LAYER	77
5.3.2	DEFECT ENGINEERING FOR ION BEAM SYNTHESIS OF SOI STRUCTURES	80
5.4	SIGE SEMICONDUCTORS AND HETEROSTRUCTURES	84
6	**CONCLUSION**	**91**
7	**REFERENCES**	**95**
8	**TABLE OF FIGURES**	**98**
9	**APPENDIX**	**99**

9.1 PHOSPHORUS AS SINTERING ACTIVATOR IN POWDER METALLURGICAL STEELS: CHARACTERIZATION OF THE DISTRIBUTION AND ITS TECHNOLOGICAL IMPACT

9.2 CHARACTERIZATION OF THE DISTRIBUTION OF THE SINTERING ACTIVATOR BORON IN POWDER METALLURGICAL STEELS WITH SIMS

9.3 SIMS INVESTIGATION OF CR – MO LOW ALLOYED STEELS SINTERED WITH BORON

9.4 2D AND 3D SIMS INVESTIGATIONS ON SINTERED STEELS

9.5 CHARACTERIZATION OF WEAR AND SURFACE REACTION LAYER FORMATION ON AEROSPACE BEARING STEEL M50 AND A NITROGEN-ALLOYED STAINLESS STEEL

9.6 SIMS INVESTIGATION OF GETTERING CENTRES PRODUCED BY PHOSPHORUS MEV ION IMPLANTATION

9.7 INVESTIGATION OF GETTERING EFFECTS IN CZ-TYPE SILICON WITH SIMS

9.8 STUDY OF DEFECT ENGINEERING IN THE INITIAL STAGE OF SIMOX PROCESSING

9.9 LOW ENERGY RBS AND SIMS ANALYSIS OF THE SIGE QUANTUM WELL

9.10 QUANTITATIVE ANALYSIS OF THE GE CONCENTRATION IN A SIGE QUANTUM WELL: COMPARISON OF LOW ENERGY RBS AND SIMS MEASUREMENTS

1 Abstract

Nowadays the application of physical analytic methods in the materials science (synthesis, characterisation and development of new materials) is essential. In this work four of these techniques, which are able to investigate surfaces and interfaces as well as the bulk material: secondary ion mass spectrometry (SIMS), Rutherford backscattering (RBS), electron microscopy (SEM, TEM) and Auger electron spectroscopy (AES); are applied on four different research areas:

1) Powder metallurgy (PM): PM is a high sophisticated technique, which enables the production of precision components with complex geometry and excellent surface quality. One important step in the part production is sintering. The enhancement of the sintering process can be done using some definite sintering additives e.g. phosphorus and boron. Here the study of the influence, pointing to the complete sintering process and to the material properties of the obtained parts, using these two sintering additives (activator), is made by means of 2D and 3D SIMS and scanning electron microscopy (SEM).

2) Tribology on the aerospace bearing materials: The formation and the effect of the reaction layer on two commonly used aerospace bearing steels (AMS 6491 M50 and AMS 5898), after two tribological tests (ball – on –disk BOD and rolling contact fatigue RCF) is investigated with SEM as well as with 2D and depth profiling SIMS.

3) Gettering effects and defect engineering: Gettering layers are produced by means of high energy ion implantation and subsequently annealing. These gettering layers (defects) are able to collect unwanted impurities and thus to reduce their concentration in the active area of the wafers, what could be essential for the further processing e.g. production of electronic devices. Copper is implanted from the backside of the wafer and thus is used as extrinsic impurity to be gettered inside these layers. Copper SIMS depth profiles show the distribution of the formed gettering layers. These defects produced by means of ion implantation can also be helpful for ion beam synthesis of silicon – on – insulator (SOI) structures. The defects can be produced prior or simultaneously with the effectively oxygen implantation. The defects and the depth profiles of all implanted species are studied by means of transmission electron microscopy (TEM), AES and SIMS depth profiles.

Abstract

4) SiGe heterostructures: The aim of this part is the comparison and correspondence of SIMS with low energy Rutherford Backscattering (RBS). The advantages and the limits of these two methods will be shown. Additionally it will be demonstrated how the mathematical simulations (RBS spectra simulation) and mathematical models and fittings (improvement of SIMS depth resolution) are able to help and to solve some problems occurring due to the limits of the analytical methods.

1.1 Abstract (in German language)

Die physikalisch – analytischen Methoden sind heutzutage in der Materialwissenschaft bei der Herstellung, Charakterisierung und Entwicklung neuer Werkstoffe und Materialien unumgänglich. In dieser Arbeit wird vor allem der Einsatz von Sekundärionen Massenspektrometrie (SIMS) ergänzt und unterstützt durch Rutherford Rückstreuspektroskopie (RBS), Elektronenmikroskopie (SEM, TEM) und Auger Elektronenspektroskopie (AES) in dem Bereich der Oberflächen - und Grenzflächenanalytik als auch in der Bulkanalysis anhand von vier komplett verschieden Forschungsgebieten gezeigt:

1) Pulvermetallurgie: Die Pulvermetallurgie ist eine hoch entwickelte Technologie, die es ermöglicht, Teile mit komplexer, präziser Geometrie und mit hochqualitativer Oberfläche in hohen Stückzahlen zu produzieren. Eines der wichtigsten Teilprozesse in der pulvermetallurgischen Herstellung der Formteile ist das Sintern. In diesem Teilabschnitt wird der Einfluss und die Auswirkungen der Sinteraktivatoren Bor und Phosphor auf den gesamten Prozess mittels 2D und 3D SIMS und Rasterelektronenmikroskopie (SEM) untersucht.

2) Tribologie an Lagerwerkstoffen: Die Bildung und die Konsequenzen einer Reaktionsschicht auf zwei verschiedenen Stählen (AMS 6491 M50 und AMS 5898), die in der Luftfahrtindustrie zum Einsatz kommen, nach den tribologischen Tests (Ball – Scheibe Test, Roll – Kontakt – Ermüdung Test) werden mit Hilfe von SIMS und SEM untersucht.

3) Gettering – Effekte und Defektproduktion (defect engineering): Mit Hilfe der Ionenimplatation und der anschließenden Temperaturbehandlung ist es möglich definierte Gettering – Schichten (Defektbereiche bzw. Defektregionen, mit denen es möglich ist nicht erwünschte Verunreinigungen innerhalb der Si Wafern aus der Aktivregion eines Wafers zu entziehen) zu produzieren. Um die Bildung dieser Gettering – Schichten nachzuweisen, wird z.B. Kupfer als Verunreinigung von der Rückseite implantiert. Dessen gemessene Tiefenverteilung dient als Abbild der produzierten Gettering – Schichten. Andererseits kann die Produktion von definierten Defekten bei der Synthese von Silizium auf Isolator (SOI) Strukturen mit der Ionenstrahlmethode behilflich sein. Die Defekte, die das ermöglichen werden sowohl vor als auch während der tatsächlichen Sauerstoffimplantation produziert. Diese Defekte und deren Auswirkungen werden mit Hilfe der Transmission

Abstract

Elektronenmikroskopie (TEM), AES und mit Hilfe der SIMS Tiefenprofile, aller an der Implantation beteiligten Spezies, studiert.

4) SiGe – Halbleiter Heterostrukturen: In diesem Abschnitt soll der Vergleich und die Korrelation der SIMS Methode mit der Rutherford Rückstreuspektroskopie (RBS, im niederenergetischen Modus) gezeigt werden. Außerdem werden die Grenzen und Vorteile der beiden Methoden diskutiert und wie man mit Hilfe von Simulationen (Darstellung von RBS Spektren) und mathematischen Modellen (Verbesserung der Tiefenauflösung der SIMS) an die Lösungen verschiedener Probleme herankommen kann, die durch die Grenzen der analytischen Methoden zu Stande kommen.

2 Introduction

2.1 General

The monitoring of each production step is the most important part in developing and producing new materials or substrates for new materials. The surface analytical techniques provide a big assistance and it is almost impossible to get on in the development and research without these techniques. Nowadays analytical requirements increase rapidly because of permanently decreasing and scaling down of the investigated devices respectively sample structures.

In this work, different applications and the results of the surface analytical techniques:

- secondary ion mass spectrometry (SIMS),
- Rutherford backscattering (RBS)
- scanning– and transmission– electron microscopy (SEM / TEM) and
- Auger electron spectroscopy (AES)

on 4 completely different research area, which will be briefly described in the following, will be presented and finally critically discussed.

2.2 Powder metallurgy

Powder metallurgy (PM), the science and technology of producing useful engineered structures beginning with production of metal powder. Powder metallurgy provides a unique opportunity to produce components with complex geometry (i.e. irregular curves) and excellent surface quality. It is suitable for single part as well as for high volume production with only little wastage of material. The production of PM components includes several steps from the raw material to the finished product as follows: powder production, mixing, forming (including pressing), sintering and if necessary after – treatment (Figure 1)[1,2]

Introduction

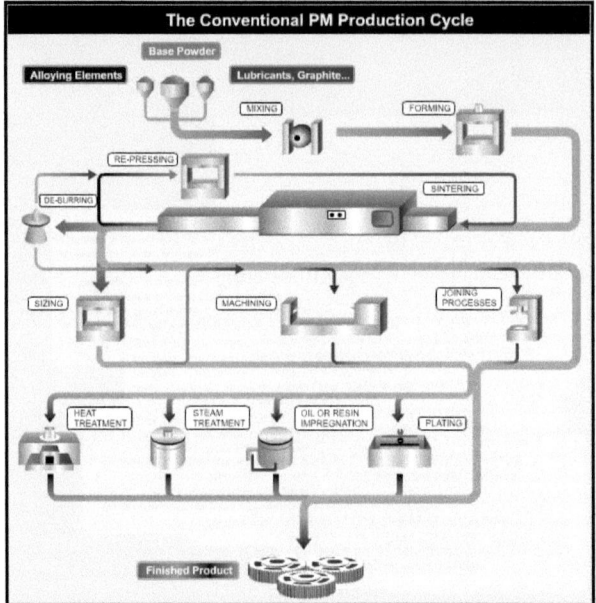

Figure 1: The PM Process beginning with powder to the end product.

The properties of the PM manufactured products are essentially the same as those of comparable cast steels, in certain cases even superior. One very important item is the porosity which can be adjusted by the amount of compaction.

The technical and commercial advantages of the powder metallurgy compared with conventional casting technology are[3]:

- The technique is particularly suited for complicated parts that are difficult to machine or for parts in which the presence of pores is essential as in filters or in self – lubricating bearings,
- Mass production of complicated parts is possible and very economical,
- Production to near net shape,
- High raw material utilization,
- No or few secondary operations,
- More uniform elemental distribution is possible because of decreased segregation problems,
- Higher micro- and macroscopic homogeneity of all compounds, yielding less waste – production and a longer life time of products.

But there are also some limitations and disadvantages of powder metallurgy:

- High cost of powder material
- The size will always change on sintering. This can usually be predicted as it depends on a number of factors including 'as-pressed' density which can be controlled
- Health problems from atmospheric contamination of the workplace
- Rapidly evolving but still not well known process

Due to the mentioned advantages, sintering technology is expanding continuously even though some applications are limited because of some materials properties (e.g. hardness) that are markedly inferior to those of wrought counterparts.

A well known powder metallurgy product is the tungsten filament for electric bulbs, but almost 70 % of powder metallurgy part production is applied at automotive engineering, for example transmission gears and connecting rods. Nevertheless the applications in the non automotive industry are increasing.

2.2.1 From powder to the end product

There are many different ways to produce iron powder for the further processes: mechanical techniques (grinding, milling), electrolytic techniques (electroplating, process where a spongy mass is deposited and can be dried and ground into powder), oxide reduction and atomization[1-4]. The investigated samples within this work were produced with "Höganäs ASC 100.29" powder[5,6]. This powder is water atomized powder and has high purity and compressibility which makes it possible to single press compacts up to densities of 7.2 g/cm^3 [6].

The base powder is blended with lubricants, alloying elements and sintering activators. Mixing operation is necessary to obtain a homogenous mixture. Lubricants are added to improve the powder's flow characteristics and to reduce friction between the powder particles and the pressing die. Zinc stearate is the most commonly used lubricant for these applications. Control over composition of the raw materials is one of major strength of the powder metallurgical process. This allows an improvement of physical and mechanical characteristics. Often used mixing aggregates are rotating drum, rotating double cone, screw mixer and blade mixer.

Mixed powder is placed into the cavity between two punches. A press squeezes powder into the shape while the compounds in the powder act as glue to hold the pressed part together (particles are bonded together by mechanical interlocking and cold welding). The cavities are usually filled at room temperature. The compaction is then pressed with a pressure between 100 and 900 MPa. The friction between the particles themselves and between particles and die wall is a major factor for the pressing. The commonly used techniques are: cold uniaxial pressing, metal injection moulding, cold isostatic pressing, hot isostatic pressing and hot forging. Hot isostatic pressing and hot forging includes sintering. The green compacts have almost size and shape (net shape or near – net shape) of the finished product but their properties are not as strong as of the finished part. Full densification by the pressing is impossible, a residual porosity is always present. This porosity could be a disadvantage because of strength decreasing, but it can also improve the lifetime of the parts. For the lifetime improvement, the uniformly distribution of the porosity is essential.

After pressing, the green compacts have the desired shape but their mechanical properties are still not satisfying. Sintering is a thermal process, which increases the strength of a green compact. The mechanism usually involves atomic transport over particle surfaces, along grain boundaries and through the particle interiors (diffusion). The thermodynamic driving force is the reduction of the specific surface area of the particles. This can be reached by densification (this could also produce shrinkage) or by reducing the pore specific surface area by forming a more spherical shape of pores. Sintering is carried out in a controlled atmosphere to protect the compact from reaction with air components (e.g. oxygen) at high temperatures. The atmosphere helps eliminating lubricants and transferring heat to the compacts. Commonly used atmospheres are: argon, nitrogen, hydrogen, mixtures of hydrogen and nitrogen, vacuum and endothermic or exothermic gases (generated from natural gas and air). All mechanical properties of green compacts are improved and frequently density increases with sintering. It is important to note that the part is not melting during the sintering process and retains its shape. The sintering temperature is below the melting point of the major component, approximately at 80% of the melting point. Binding of powder particles provides the strength of the part and heat allows the alloying additives to diffuse throughout the part. The mechanical bonds produced by pressing change to the much stronger metallic bonds[7].

As mentioned above the sintering temperature is at about 80% of the melting point of the major component. Two of the important tasks of the sintering activators are decreasing of the sintering temperature and sintering time. Sintering activator enhance sintering by the formation of liquid phase. The formation of the liquid phase has an effect on the final density and on the microstructure[8, 9, 10]. During the sintering process added lubricants leave the part because of the increasing temperature (e.g.: zinc stearate decomposes at ~ 300°C). Many parts are finished at this stage, some will require additional processing.

Sintering can be enhanced by the presence of a liquid phase[1]. During the sintering the liquid phase can be reached by the addition of alloy elements or sintering activators. A liquid phase is formed when the sintering temperature is between the melting points of the added and major component, by the melting of eutectic phase mixtures formed by diffusion, or by incipient melting. Transient (temporary) liquid phase sintering is characterized by the formation and disappearance of a liquid phase during sintering. The liquid flows between the powder particles, fills pores and causes densification. The liquid phase provides a better distribution of the alloying components, a spheroidisation of the pores, enhances diffusion and reduces the sintering time. It also causes shrinkage. Some liquids even attack the sintering material[2, 4].

After sintering the parts may go through a series of secondary operations to improve sintered part's material strength, density, appearance, and performance or make sure components fall within required tolerances. These include[11]:

- *Repressing*: After initial sintering, some parts may be pressed again and resintered. Typically, this step is applied to achieve a higher density. It is also done when dimensional tolerances are extremely tight, to press the part much closer to its final shape before final sintering.
- *Sizing*: Sizing is repressing a part after all completed sintering operations. It typically involves deforming of very small amount of material.
- *Machining*: Machining is done if certain part geometries are required.
- *Joining*: For those parts that cannot be joined during the sintering process. More conventional options are: welding, brazing or fastening by mechanical means (such as with screws or bolts).
- *Plating*: If special surface properties such as corrosion resistance or wear resistance are required the parts can be plated using any of a number of processes (black oxides, phosphate coatings or zinc plating are few possibilities).

- *Impregnation*: The parts with pores can be impregnated with oil or resin. Impregnation provides a unique possibility of altering the wear or surface characteristics of the finished parts.
- *Steam treatment*: To increase corrosion resistance of the parts, steam treatment is often used instead of more expensive plating and impregnating processes.
- *Heat treatment*: When special hardness characteristics are required the parts can undergo heat treatment.

2.2.2 Topics of investigation

The use of phosphorus and boron, acting as sintering activator, was investigated depending on some parameters: the present concentration of the used activator, sintering time, sintering atmosphere, sintering temperature and alloying elements. The investigations were concentrated on the secondary ion mass spectrometry (SIMS) 2D and 3D element imaging of the sintering activators as well as of other alloying and trace elements in order to obtain how these elements affect the mechanical properties of the investigated samples and how the sintering process is influenced by the addition of these elements. Auger electron spectroscopy (AES) was applied to detect the grain boundary segregation of phosphorus and scanning electron microscopy (SEM) was used to identify the fractures obtained by means of the Charpy impact tester. The mechanical and physical properties of the obtained material will be described on the basis of the impact toughness, fractography, density and hardness measurements.

2.3 Tribology

2.3.1 General

Tribology (*Greek, tribos...friction*) is defined as the science and technology of interacting surfaces in relative motion, and involves the study of friction, wear and lubrication and thereby incorporated interface interactions between solids, liquids and gaseous[12]. Among many application areas:

Introduction

increasing performance, improvement of system effectiveness, reduction of energy consumption by decrease of friction, increase of product reliability, lifetime improvement and decrease of maintenance and servicing costs of used materials, the study of tribology is mostly applied in bearing design. The overall aim of friction study is using its knowledge to save energy, environment and resources due to reduction of friction inside the tribological systems and development of biological lubricants.

A first decrease of friction or wear can be achieved by the change of material composition for example by means of additional hardening or thin film deposition on the material surface. The other way is the usage of lubricating oils (synthetic or mineral oils) as well as of molybdenum disulphide or graphite.

The maximum performance of different material systems can be reached by optimizing dynamics, strength and stiffness of all system components. Today the material requirements on new developed materials for e.g. engine main shaft bearings, increases strongly due to an elevated speed index (bearing bore diameter multiplied with rational shaft speed) and slip ratios[13]. Generally the materials e.g. steel is protected by the reaction with an additive, which forms a layer between surfaces which are permanently in contact[14]. The reaction layer formation between these two in contact surfaces is strongly affected by the tribological loadings conditions. Further on it depends on the materials used, lubricant and the working temperature. The lifetime of materials can be enhanced by the decrease of wear and friction. This is available by an appropriate balance between reactivity and of materials and lubricant used in tribological systems.

The choice of materials is critical in the design of bearings. The correct choice of materials should maximize fatigue life, and specify corrosion resistance. The environmental conditions must be carefully considered and evaluated in order to make the correct material decisions. Bearings in aircraft machines operate at high speeds and adverse environmental conditions and need to perform faultless. The standard steels for bearings for aircraft engines are AMS 6491 (M50) and AMS 6278 (M50NiL). The traditional M50 steels are commonly heat treated to achieve optimum hardness and dimensional stability, and thus those are suitable for most operating conditions. Both, AMS 6491 (M50) and AMS 6278 (M50NiL) steels have a chromium content of about 4.1 wt%. This very low chromium content causes a worse corrosion resistance due to the surface of the materials is covered with iron oxides mainly[15]. Mobil Jet II is the oil mostly used in the aircraft engine bearings in the recent years[16]. It contains tricresyl phosphate (TCP[16]) which is adsorbed by means of chemisorption to the operating surfaces at about 200°C and thus reduces wear and friction at these operating temperatures[17]. Investigations have shown that TCP reacts with different iron oxides, also with the iron oxides covering the AMS 6491 and AMS 6278 but only hardly with chromium oxide[18]. During the last years new steels were developed based upon higher chromium content inside the matrix in order to improve the corrosion resistance of the material. One of these new developed high strength stainless steel grade, which was the investigating material within this work was AMS 5898[19, 20, 21] whereby the improved corrosion resistance is the result of the surface formation of protective passive layer containing chromium oxide.

2.3.2 Tribology testing

Among many tribological tests, the samples which were investigated within this work were checked by means of rolling contact fatigue test (RCF) and ball – on – disk (BOD) test.

2.3.2.1 Rolling contact fatigue (RCF)

Generally, rolling contact fatigue describes the damage of material after repeated rolling respectively sliding contact. The typical application areas where the rolling contact fatigue is studied are bearings and bearing materials as well as rails. The rolling contact fatigue is influenced by high contact stresses, plastic deformation, exhaustion of ductility and cracks initiation (which can lead to the failure). There are many types of devices investigating the rolling contact fatigue: four ball machine, five ball machine, ball on plate machine, ball on rod machine, disk on rod machine and contacting ring machine. The tests made within this work were performed by means of the modified three balls on rod machine. Figure 2 shows three (steel) balls, orbit a rotating cylindrical sample (rod) which is direct driven by electric motor. The thrust load is applied mechanical and the lubricant is supplied by constant dripping onto the top of the rod[22, 23].

Introduction

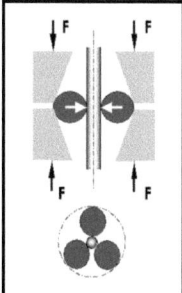

Figure 2: Schematic illustration (loading configuration) of the three balls on rod RCF test (side view on top and top view at bottom). The arrows mark the load direction.

2.3.2.2 Ball – On – Disk (BOD)

For the determination of the friction coefficient a test body is pressed with a defined force and it is sliding over a counterpart substrate. The force (friction force) which is needed to ensure the constant sliding speed is determined. For ball – on – disk test a ball is loaded to sample (disk) with a well known force. The ball is mounted on a stiff lever, which is designed as a frictionless force transducer. The friction coefficient can be determined during testing by measuring the elastic arm deflection. A schematic overview of a CSM tribometer is shown in Figure 3[24].

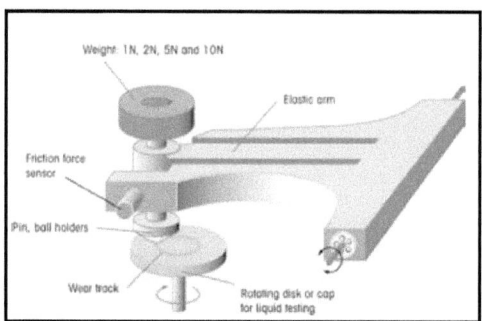

Figure 3: Schematic overview of a CSM tribometer[24].

During the continuously test the penetration depth of the test body gets higher and the friction contact surface increases. Normally, this induces a higher frictional resistance and complementary a higher friction coefficient which is demonstrated in the Equation 1:

Introduction

$$\mu = \frac{F_F}{F_N} \qquad \text{Equation 1}$$

μ friction coefficient
F_F friction force
F_N normal force (load)

Knowing the friction coefficient, the information of the tribological system e.g. dry – sliding (if no lubricant is used), abrasive wear properties and hardness can be obtained. The wear coefficient can be determined by the volume material lost during the test. The analysis of the testing partner is made by means of the optical profilometer (3D imagination of wear) as it is demonstrated in the Figure 4.

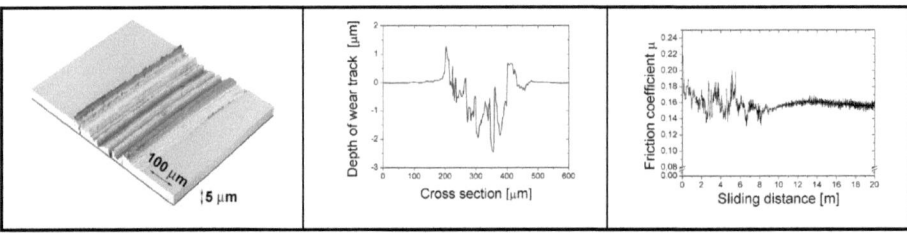

Figure 4: (left) Wear track image of disk made by optical profilometer after BOD test; (middle) corresponding cross section and (right) friction coefficient.

The BOD tests have to be carefully conducted because the friction behaviour strongly depends on environmental parameters such as temperature and humidity. Beside the BOD test other geometries of the static partner are also possible e.g. pin mounted on the testing instrument (pin – on – disk test, POD).

2.3.3 Topics of investigation

The reaction layer formation on surfaces in contact was studied on two different materials, standard aerospace bearing steel AMS 6491 (M50) and high strength stainless steel grade AMS 5898 after rolling contact fatigue tests (contact pressure: 6 GPa) and ball-on-disk tests (contact pressure: 1.6 GPa, sliding speed 10 cm/s) at room temperature and at 150°C. The contact areas were investigated after the tests by means of optical profilometer, scanning electron microscopy (SEM) supported by secondary ion mass spectrometry (SIMS) in order to determine the element distribution within, type of -, thickness of - and homogeneity of the formed reaction layer.

2.4 Ion implantation and defect engineering for semiconductors

2.4.1 General

Generally, semiconductors are materials, whose electrical conductivity at room temperature is between those of highly conducting metals and poorly conducting insulators. Pure materials, whose electrical conductivity is determined by their inherent conductive properties, are called intrinsic semiconductors. Typically representatives of intrinsic semiconductors are pure elements silicon and germanium. These are the elements of the IVA group in the periodic table of elements and have diamond cubic structure (Figure 5) with highly directional covalent bonds.

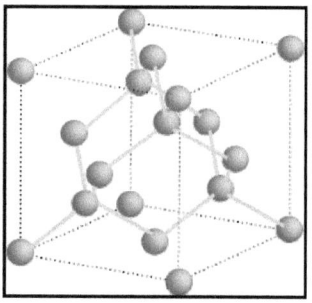

Figure 5: Diamond crystal structure.

Introduction

For silicon or germanium the bonding electrons are unable to move through the crystal lattice and therefore to conduct electricity unless sufficient energy is provided to excite them from their bonding position. Afterwards the electron becomes a free conduction electron and leaves behind a positively charged "hole" in the above mentioned crystal lattice. For the electrical conduction process for intrinsic semiconductors (pure silicon or germanium), both electrons and holes are charge carriers in an applied electric field.

The two commonly used models for the description of the conductivity are electron – hole pair generation model and the energy band theory. The model of electron – hole pair generation[25] explains the higher conductivity at higher temperatures. The lattice vibrations are much stronger at rising temperature and thus more bonds are broken and charge carrier concentration is increased. The energy band diagram[26] for the intrinsic semiconductors is described by the electron occupation of the levels in the lower valence band, which is almost filled at room temperature. Between the valence band and the above located, almost empty, conduction band, the forbidden energy band gap is situated. The energy band gap is 1.1 eV for silicon at 20°C[27] and there are no energy states allowed in it. At this temperature (20°C) the thermal energy is sufficient to excite some electrons from the valence band into the conduction band leaving vacant sites or holes in the valence band, what exactly means that two charge carriers are created.

With increasing temperature there is a difference in the conductivity behaviour of metals and semiconductors. With increasing temperature the conductivity of metals is decreased. In opposite with increasing temperature the conductivity of semiconductors is increased. The number of electrons with sufficient thermal energy to enter the conduction band is proportional to $\exp(-E_g / 2kT)$ whereby E_g is the thermal energy gap between valence and conduction band, k the Boltzmann's constant and T the temperature.

Doping (adding "impurities" to material) semiconductors lead to dramatically changes in its electrical properties. This new kind of semiconductor is termed extrinsic semiconductor. Usually the concentration of the doping element is in the range of 10 to 1000 ppm. Adding atoms of the V group (5 valence atoms) e.g. phosphorus, arsenic or antimony. The introduction of excess electrons, which are only loosely bonded to the positively charged phosphorus nucleus, occurs here and these electrons can migrate from the valence to the conduction band. Semiconductors doped with the elements of the V group are called n – type semiconductor (negative – carrier type) due to the concentration of the electrons is higher then the concentration of holes (Figure 6).

Doping can also be provided with elements of the III group (3 valence atoms) e.g. boron, aluminium or gallium. Hereby the electrons can easily migrate to the new energy level (acceptor level) generated by the impurity, which is higher then the valence band level. Thereby holes are

Introduction

generated in the valence band which act as positive charge carrier. Semiconductors doped with the elements of the III group are called p – type semiconductor (positive charge carrier) due to the concentration of the holes is higher then the concentration of electrons (Figure 6).

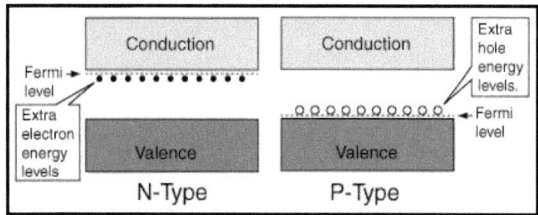

Figure 6: Energy band diagram for n – and p – type semiconductors (silicon)

Widely used semiconductor devices are diodes, bipolar junction transistor, field electron transistor (metal oxide semiconductor field effect transistor or MOSFET), complementary metal oxide semiconductor (CMOS) and bipolar CMOS.

2.4.2 Ion implantation

The ionic bombardment of semiconductors was patented in the 1954 by William Schockley at Bell Laboratories[28]. It took then the next 15 years to develop the method of the controlled ion implantation into materials and with rising development of semiconductor industry the ion implantation became the method of choice for the introduction of dopants into materials in particular into semiconductor. The scheme overview of an ion implantation device is given in Figure 7.

Introduction

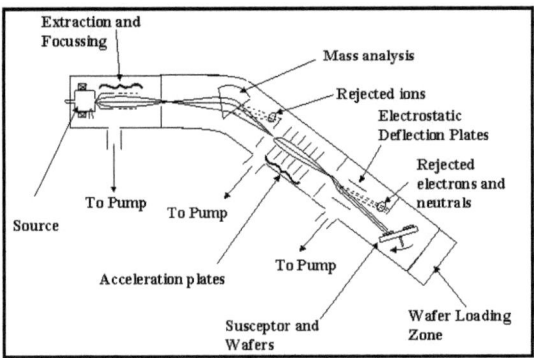

Figure 7: Schematic overview of a ion implanter tool[29].

The implantation atoms are vaporized and ionised, generating a focused ion beam, accelerated with well defined energy, pass the mass analysis tool (magnet prism) and directed towards the substrate. The ion implantation process is mainly determined by implanting species, implantation energy and the implantation dose. Impacting the substrate, the atoms enter the crystal lattice, collide with substrate atoms and lose gradually their kinetic energy and stop finally at the definite projection range ($R_P \pm \Delta R_P$). The mean projected range R_P can be manipulated by adjusting the acceleration energy and the ion fluence can be monitored by the ion current during the implantation. Using the simulation software programs and by knowledge of the process parameters it is possible to predict the mean projected range of the implanting species. Two of such simulation software programs are described in the chapter 2.4.3. The real state of the projection range and the distribution of the implanted species can be extracted using the depth profiles made by means of secondary ion mass spectrometry (SIMS).

Impacting the substrate each implantation ion displaces atoms inside the substrate, thus damage (point defects) along the projection range is caused. This residual damage can be healed by a moderate heat treatment process (annealing) but it acts also as an effective gettering centre for impurities like transition metals in silicon and will be described in the chapter 2.4.4. Furthermore ion implantation can be used for introduction of impurities through the certain surface layers and also to introduce impurities which are difficult to be introduced by other processes (e.g. CVD).

2.4.3 SRIM simulation

The Stopping and Range of Ions in Matter (SRIM)[30] is a simulation program calculating the stopping and range of ions with the energy range of 10 eV to 2 GeV / amu into matter using quantum mechanical treatment of ion (note for later usage: moving respectively impacting ion) – atom (note for later usage: target respectively stationary atoms) collisions. Impacting the target the ion and atom have screened Coulomb collision, including exchange and correlation interactions between the overlapping electron shells. Electron excitations and surface plasmons (collective excitation of *free* surface electrons) are also created by the long range interaction of ions with target atoms. The concept of effective charge, including a velocity dependent charge state and long range screening due to the collective electron sea of the target, describes the charge state of the ion within the target[30].

The transport of Ions in Matter (TRIM) is one of the most used comprehensive programs included into the SRIM software package[31]. It bases upon a Monte – Carlo calculations, which follows the ion into the target, making detailed calculations of the energy transferred to every target atom during the collision. It allows the use of complex targets e.g. consisting of eight layers each with different material. It calculates both the final 3D distribution of the implanted ions and also all kinetic phenomena associated with the loss of the ion's energy: target damage, sputtering, ionization, and phonon production. The calculation is made very efficient by the use of statistical algorithms. Target atom cascades within the target are followed in detail. One example of oxygen impaction (200 keV) into silicon is demonstrated in the Figure 8.

Introduction

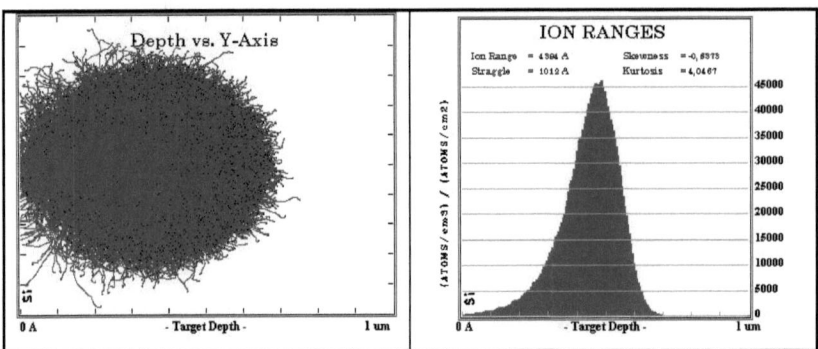

Figure 8: Result of the TRIM calculation of oxygen impaction (200 keV) into silicon: collision plot showing all trajectories of 100000 calculated ions (left) and ion distribution along the depth axis (right).

The dynamic version of TRIM is called T - Dyn[32, 33]. It considers the penetration of the primary ions into the target, regarding on both nuclear scattering and the electronic stopping of the ions. Further on, the development of atomic collision cascades, which is responsible at first for the atomic mixing and second for the sputtering of the target material and the modification of the atomic composition of near surface region of the target material due to the ion implantation, is also included into T – Dyn. Because of the interactions of the primary ions with target atoms as well inter atomic collisions in the cascades the introduction of the interaction cross sections for the nuclear scattering and electronic stopping was done by Ziegler – Biersack – Litmark[34].

Finally, the simulation – and calculation – software (and results) are essential tools at the semiconductor especially at the implantation research and industry.

2.4.4 Gettering layers (R_P, $R_P/2$, trans R_P) and defect engineering

As mentioned above the substrate is damaged in its crystalline structure during the ion implantation. With a moderate heat treatment (annealing) this damage can be partly healed by point defect recombination. The residual damage can act as an effective gettering centre for impurities like transition metals in silicon devices. Such impurities can strongly influence, in particularly degrade, the properties of silicon devices[35]. The answer to the problem is the gettering of impurities, whereby the unwanted metal impurities are collected and their concentration in the device active area is reduced, also known as proximity gettering[36]. High energy ion implantation in the MeV range can be applied to getter the metal impurities in a buried layer slightly deeper than the device region.

A typical high energy (MeV) implantation causes a displacement of 10^3 atoms along the trajectory line of each impacting ion. Hereby each atom displacement results in one self – interstitial and in one vacancy (Frenkel pairs). The investigations have shown that gettering layer is not only formed around the mean projected ion range (R_P), but also in the region between surface and the R_P. This is termed $R_P/2$ – effect. The radiation induced vacancies and interstitials are assumed to recombine locally during annealing. This process leads to a spatial separation of the generated vacancies and self-interstitials resulting on average to a vacancy – rich region at Rp/2 and an excess of interstitials in the Rp region. With exception of the implanted atoms (+1 atoms) only the point defects remain which are in (local) excess[37, 38, 39, 40, 41, 42, 43, 44]. The gettering sites for impurities around Rp/2 are ascribed to the excess vacancies and the gettering sites at Rp to excess interstitials. The excess interstitials around Rp form interstitial loops and dislocations during annealing which can be easily observed by cross section transmission electron microscopy (XTEM)[41, 43, 45]. Further on the third gettering layer was detected to be in the regions beyond the mean projected range and it was only detected for implantation species like P^+ or As^+ and is termed trans – R_P – effect. It is assumed that the gettering centres in the trans – R_P region are interstitial cluster formed during annealing by implant atoms and point defect diffusion. The gettering centres in the trans-Rp region are not yet detected by TEM[41, 46, 47].

2.4.5 Defect engineering for ion beam synthesis

Defect engineering can also be applied for in situ generation of defects during the implantations e.g. for ion beam synthesis of silicon on insulator (SOI) structures. SOI semiconductor technology provides higher performance (15% faster, 20% less power) devices than traditional silicon technique. A typical SOI wafer is described by a thin silicon oxide or glass layer (80nm) between the silicon substrate and cap silicon layer (50nm – 1µm). The devices are situated into the cap Si layer and this arrangement reduces the amount of the electrical charge that e.g. transistor devices have to move during the switch operation. The devices are able to function at significantly higher speeds while reducing electrical losses. Nevertheless SOI devices are more expensive than traditional silicon devices, thus these are predominantly for high – end applications (e.g. portable computing devices) used.

Introduction

The SOI wafers are commonly produced on three different ways:
- Wafer bonding: two wafers, both coated with an insulating layer (oxide), are bonded with the insulating side together in a furnace forming one new wafer with a buried oxide layer (BOX) between two layers of semi conducting material. The new material is then lapped and polished until the desired thickness of the cap layer is reached
- Smart cut: The technique is similar to wafer bonding, but here the excess of semiconductor on the future device side is removed by hydrogen implantation to an estimated depth and subsequently annealing at 500°C which is responsible for the split along the introduced stress plane. Here the excess on semi conducting material can be used for further preparation of SOI wafers.
- Separation by Implantation of Oxide (SIMOX): Hereby oxygen is introduced into silicon by means of ion implantation. The subsequently annealing creates a buried oxide layer (BOX). The ion implantation parameters such as ion fluence and the implantation energy define the form and the placement of the BOX.

The last described technique (SIMOX) can be coupled with preliminary described defect engineering for ion beam synthesis of SOI structures. The ion beam synthesis is the technique that provides the creation of new domains inside one matrix by specific implantation and after treating process parameter (ion dose, implantation temperature, simultaneous dual beam implantation)[48]. The creation of defects prior, or during the implantation process and the introduction of the wanted implanted atoms itself can produce significant modifications of the properties of the substrate material. Typical ways here are pre deposition of cavities, e.g. by means of He pre implantation, or by simultaneous implantation of oxygen and silicon creating in – situ vacancies.

2.4.6 Topics of investigation

In the first part, the differences in the impurity gettering behaviour were investigated depending on the implanted ions (P^+ and Si^+), implantation dose and annealing time at 900°C. Secondary ion mass spectrometry (SIMS) depth profiles were recorded of extrinsic impurities copper, and intrinsic impurities oxygen and carbon to obtain information about their gettering behaviour. It will be shown that the impurities are gettered at the mean projected range (Rp) of implanted ions, Rp –

effect, at around half of the projected ion range, Rp/2 – effect and even at permitting circumstances impurities were gettered beyond Rp, trans – Rp – effect.

In the second part of this work SIMS supported by transmission electron microscopy TEM was primarily applied to detect the in depth distribution of oxygen. The overall aim here was the application of defect engineering methods for the reduction of the oxygen ion dose and the improvement of the crystal quality in the top Si layer at the ion beam synthesis of silicon on insulator (SOI) structures.

2.5 $Si_{1-x}Ge_x$ heterostructures

2.5.1 General

The semiconductor high tech industry is one of the most growing branches. As already described above the main semi conducting material is silicon followed by germanium and gallium – arsenide doped with elements like boron or phosphorus. The research and the development of new materials and devices as well as the improvement of the material and electronic properties of those are the main areas of the semiconductor research industry. In generally the device size of the new developments is always scaled down which is always a new challenge for the existing analytical methods.

One possible way for new development is the combination of base semi conducting materials. Among these relatively new developed materials are SiGe heterostructures and quantum well rich materials. Here materials are coupled with different band gap energies. On the example of $Si_{1-x}Ge_x$ heterostructures $Si_{0.75}Ge_{0.25}$ is the material with lower band gap energy and $Si_{1-x}Ge_x$ (x = 0...0.20) respectively the material with higher band gap energy. In case of quantum well material here Si would be the material with higher band gap energy and $Si_{1-x}Ge_x$ respectively the material with lower band gap energy[49]. Using heterostructures the specific manipulation of electrons and holes is possible. This procedure is termed as band gap engineering and the obtained devices based on these materials provides extremely interesting electronic (pointed to transport) and optical (pointed to luminescence) properties. By doping the barrier of a quantum well with donor impurity dopants, a two – dimensional electron gas (2DEG) can be formed. This system has interesting properties at low temperature, exhibiting the quantum Hall effect[25]. Figure 9 shows the typical design of Si_{1-}

Introduction

$_x$Ge$_x$ heterostructures whereby the red line is the new engineered conduction band variation and the blue line is the wave function of the first state.

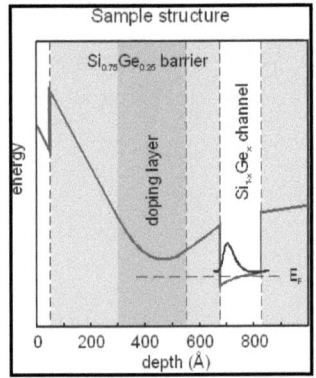

Figure 9: Typical composition of Si$_{1-x}$Ge$_x$ heterostructures.

Due to their quasi two dimensional nature, the electrons in quantum wells have sharper density of states than those of bulk material and thus the typical microelectronic devices based on these materials are heterojunction bipolar transistors (HBT)[50]. HBTs are used in low noise electronic applications and have significantly higher forward gain and lower reverse gain which translates into far better low current and high frequency performance than available from commonly used homojunction or traditional bipolar transistors. Further devices are resonant tunnelling diodes (RTD), high electron mobility transistors (HEMT). The production of quantum cascade lasers (near – to far – infrared wavelength ranges and semiconductor diode laser) bases also on these materials[51].

However it is important for devices based on these materials that the interfaces between the structures within the material have abrupt transition that means that the composition is changed within a few atomic layers. In generally the heterostructures as described above can be produced by means of metal organic chemical vapour deposition (MOVCD) or molecular beam epitaxy (MBA). Using MOCVD the semi conducting materials are deposited from the flowing gas vapour over the heated substrate. An example of the GaAs production is shown in the Figure 10.

Introduction

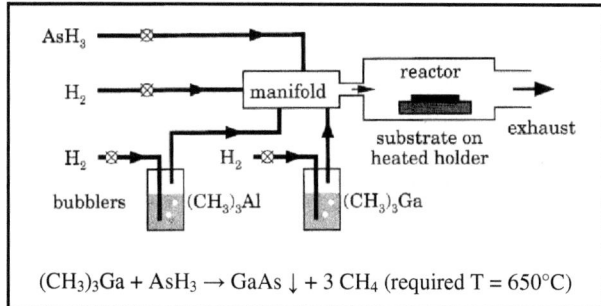

Figure 10: MOCVD production and deposition of GaAs[52].

The principle of MBA will be described more in detail in the following chapter.

2.5.2 Molecular beam epitaxy (MBE)

Molecular beam epitaxy (MBE) is one of the highly developed techniques for growth of very thin layers. It has been developed in the early '70s[53, 54].

It is possible to produce very thin (monolayer range) but also very complex layer with well defined and abrupt interfaces with atomic layer precision. In one ultra high vacuum chamber (supported by a system of cryopumps, and cryopanels, cooled down using liquid nitrogen to a temperature of -196° Celsius) the beam of atoms or molecules is directed towards the substrate where subsequently a crystalline layer is formed. MBE equipment consists of three different chambers: a growth chamber, a buffer chamber and a load lock. The load lock is necessary to bring the substrates and the finished samples inside and out of the vacuum system. The buffer chamber is used for the pre – preparation and storage of the samples. The operating pressure is about 10^{-10} Torr and the operating temperature can be modified into the range of 1500°C. The very big advantage compared with the conventional evaporating techniques is very low deposition rate, whereby every impinging atom is able to migrate over the substrate and find his place to build up a new crystal lattice. The deposition rates are in the range of 0.X Å/s, controlled over the temperature material flux dependence. An automated computer and shutter system ensures the precise control of the thickness of each layer. This assures the controlled epitaxial growth[55] and therefore the formation of very abrupt interfaces is possible. Hereby the material structures, where the electrons can be confined in space, producing quantum wells or even quantum dots, can be obtained. The in situ analysis of the growing films is

Introduction

possible e.g. by means of Reflection High Energy Electron Diffraction (RHEED). A schematic overview of MBE equipment is given in the Figure 11.

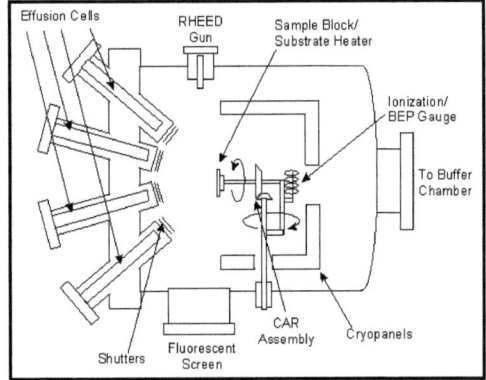

Figure 11: Schematic view of an MBE growth chamber[56].

2.5.3 Topics of investigation

The MBE grown SiGe samples were investigated on the germanium concentration inside, the position, the thickness of the quantum well by means of Rutherford backscattering (RBS) and secondary ion mass spectrometry (SIMS). The aim in this part was to compare the result obtained by means of these two methods and to demonstrate their opportunity for usage in semiconductor and microelectronic industry. Furthermore it will be shown that the use of mathematical models e.g. forward convolution (point to point convolution) is very helpful to improve the depth resolution of the used SIMS instrument. Nevertheless the detection of trace elements (here doping element antimony) by means of SIMS, which could not be evaluated with RBS in low energy mode, will be demonstrated.

3 Analytical Techniques

Among the many differentiation possibilities, the classification of the analytical techniques using the primary excitation is the widely used. The analytical techniques used throughout this work can be divided into ion beam – excitation (Figure 12) and electron – excitation (Figure 13) techniques.

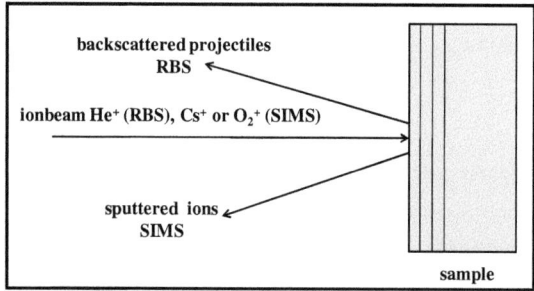

Figure 12: Schematic overview of the ion probe techniques: measuring backscattered projectiles, Rutherford BackScattering (RBS), and sputtered ions, Secondary Ion Mass Spectrometry (SIMS).

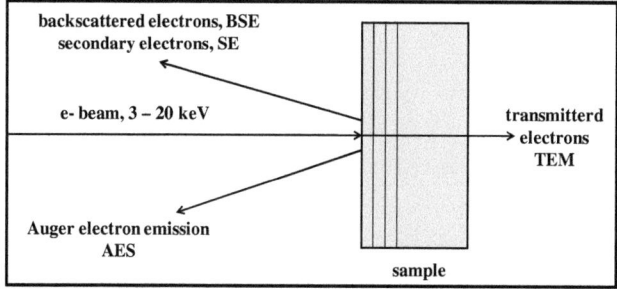

Figure 13: Schematic overview of the electron probe techniques: collecting Auger electrons, Auger Electron Spectroscopy (AES), measuring (imaging) of the BackScattered (BSE), Secondary (SE) and Transmitted Electrons (TEM).

Depending on the used technique it is possible to determine the chemical, elemental and in some times molecular composition of material surfaces and interfaces as well as the bulk composition. Generally it is at first not important, which method will be used, but it is well important that information about the sample are reached quick, cheap and accurate. Ion beam techniques are in use, when the information cannot be obtained with other techniques and these are generally expensive and also difficult to care. Nevertheless the ion beam techniques are quantitative, sensitive, almost no sample preparation is needed, in general quick but some damage the samples (SIMS).

The results that will be presented in this work are based upon the results made by myself by means of SIMS supported by RBS, AES and SEM. That's why in the following SIMS technique will be described more in detail and for AES, RBS and SEM a typical overview will be given.

3.1 Secondary Ion Mass Spectrometry – SIMS[57, 58]

Secondary Ion Mass Spectrometry (SIMS) is a sputter based analytical method. It provides the ability to detect all elements with a high analytical sensitivity, good lateral and depth resolution. Further the images gained during profiling can be stored and processed afterwards. The principle of SIMS was introduced in 1949 by R. Viehböck und F. Herzog. Since than, many modifications and improvements have been made[59].

3.1.1 Principle of SIMS

An ion beam is used for sputtering (Figure 14) and ions emit from the surface, which are called secondary ions. They are detected with a mass spectrometer. The incident ions, deposit their energy into a specimen through processes such as cascade collision, backscattering and recoil implantation. These processes happen in an average depth of 10 - 20 atomic layers dependent on the energy, the impact angle and the mass and atomic number of both primary ions and target atoms. The energy transfer causes an impulse transmission, which leads to an expulsion of particles of the first 1 - 3 atomic layers.

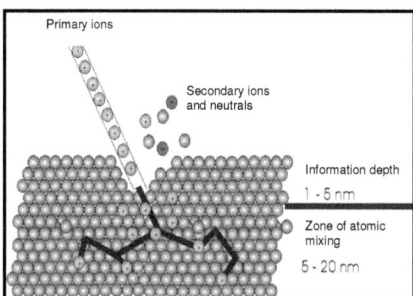

Figure 14: The sputter process.

In common, Cs^+, O_2^+, O^-, Ar^+ and Ga^+ serve as primary ions. The energy of the primary ions ranges from 0.2 keV up to 30 keV. The amount of particles driven out per incident primary ion is called sputter yield and comes up to 0.1 - 10 particles per primary ion. Most of the particles are uncharged. Only 10^{-3} % - 1 % are ionized and can be extracted to the mass spectrometer. These ions are not just singly charged atom ions but also singly and multiple charged molecular and cluster ions. Fortunately, molecular and cluster ions differ from atom ions in their kinetic energy so they can be prohibited from entering the mass spectrometer by energy filtering. The rest of the sputtered species consists of neutral atoms and clusters.

There are several ways of analyzing the ions by their mass to charge ratio. The most common mass spectrometers used in SIMS analyses are double focusing sector field instruments, quadrupole instruments and, especially for pulsed SIMS, time of flight mass spectrometers. The advantages of the double focusing sector field devices are the high mass resolution and stigmatic properties. Quadrupole instruments have a high mass scan speed and time of flight mass spectrometers are very flexible. The detection of the secondary ions is done with an electron multiplier, a faraday cup or, for stigmatic applications, with a channel plate coupled with a fluorescence screen.

3.1.2 Operating modes

3.1.2.1 Imaging mode

In imaging mode (Figure 15) all pixels (picture element) of the element distribution are detected simultaneously by the use of a stigmatic secondary ion optic. The lateral resolution of the image is not primary influenced by the primary ion beam diameter. The resolution is determined by the energy distribution of the secondary ions, the energy (and angle) distribution results in a chromatic aberration of the first electrostatic lens, the emission lens. The resolution can only be enlarged by limiting the energy (or angle) acceptance of the instrument, which of course reduces the transmission and therefore the detection limit. In practice this effect limits the lateral resolution to about 1 μm, only for main components 0.5 μm are possible.

In practice the quality of the image is also reduced by the use of high mass resolution and energy offset, therefore it is in many cases not possible to avoid mass interferences. By the use of image processing tools for classification of mappings of different masses a determination of element distribution is possible.

For secondary ion detection a lateral resolved detector is necessary. In the first step a channel plate for amplification is used, the secondary electrons of the output of this device are then accelerated either to a fluorescent screen or to a resistive anode. In the case of fluorescent screen the image is picked up by a CCD (coupled charged device) camera and summed up frame by frame by computer software. This system has the advantage of unlimited secondary ion intensities, but compared to the digital detection of the resistive anode decoder the lateral and intensity linearity is not so good defined.

Analytical Techniques

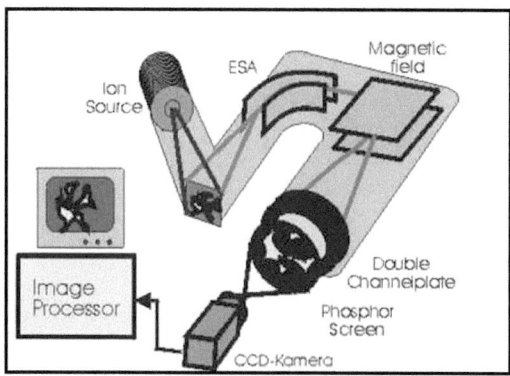

Figure 15: SIMS in imaging mode.

3.1.2.2 Three dimensional (3D) imaging

3D SIMS is a further development combining dept profiling (described in the next chapter) and imaging analysis. In imaging mode the summing up time of single ion distribution is low (approximately 5 – 7 s). This enables the possibility to cyclic image acquisition during a normal depth profile. Measurement of large and therefore representative volumes (150 x 150 x D µm³, D....depth....5 – 10 µm) in short time (~ 1 h) are possible.

In scanning mode (described in the next chapter) the sequential detection of the single pixels (picture elements) and volume elements results in long measuring times, therefore in practice only small volumes (10 x 10 x 0.4 µm³) can be measured.

The advantage of imaging mode is the fast data acquirement. Due to the fact that all pixels are projected and detected simultaneously, the measurement time of one element distribution is significant low.

3.1.2.3 Depth profiling mode

Depth profiles are recorded with a more focused beam compared to that of imaging mode. A large area in common between 1 µm and 400 µm is scanned fast. This area is referred to as scanned area. In order to prevent crater edge effects, the ions analyzed escape from a smaller area in the centre of the scanned area, called analyzed area. The size of the analyzed area is normally determined by the size of the extraction lens, which accelerates the secondary ions into the mass spectrometer. When the primary ion beam is focused to a diameter below the size of the extraction lens and the illuminated area is smaller than the extraction lens, the analyzed area is determined by the illuminated area, which is at least the diameter of the beam.

3.1.2.4 Raster scan mode

A very fine focused primary ion beam with a diameter typically less than 1 µm scans the surface of the sample from point to point. The secondary ions are detected with an electron multiplier and the mass spectra of the points are fit together to obtain an image of the surface. Usually the beam intensities are very low. So depth profiles take too much time and counting statistics become worse, because the secondary ion intensity is very low. In order to improve the counting statistics, the counting time of the electron multiplier can be raised. This additionally extends the time of measurement. The advantage of this mode is the good lateral resolution, which is determined through the diameter of the beam. Care has to be taken to optimize the beam both in diameter and homogeneity.

3.1.3 Analytical characteristics

At the beginning of this chapter a general idea of the characteristic analytical features of SIMS analyses is provided. Following, the attributes will be discussed.

The most important analytical features are:
- low detection limits
- isotope specificity
- ability to detect all elements
- surface sensitivity
- high depth resolution

- quantification of elements with higher concentrations very difficult
- little accuracy in quantification
- standards needed for quantification
- isolators are difficult to measure
- only samples that withstand high vacuum conditions are measurable
- samples must fit into the sample chamber
- no non-destructive method
- mass spectra are hard to interpret

The detection limit ranges from 10 µg/g for insensitive elements to 0,01 ng/g for sensitive elements[60]. The sensitivity of an element can be improved by means of reactive sputtering. Reactive sputtering means enhancing the ionization probability of an element using primary ions, which have huge electro negativity for electropositive elements or lower electro negativity for highly electronegative elements.

Isotope specific characterization and the possibility to detect all elements are general features of mass spectrometry. The limits depend on the mass resolution of the mass spectrometer, which is defined as $\Delta m / m$. Usually the mass resolution lies between 300 and 10000.

Surface analyses are a tricky task anyway. Bombarding the specimen with ions gives rise to additional problems. The ions destroy the structure of the sample and are implanted. This causes enhanced segregation and diffusion. Moreover, the primary ions push the atoms of the sample forward into the sample, which causes broadening of the signal. Additionally the composition of the sample is modified. The advantage of SIMS is that the particles detected come from the first three atom layers in the sample. An additional problem arises when the sample surface is rough.

The depth resolution depends on the impact angle, the energy of the primary ions, the mass and atomic number of both the primary ions and the atoms of the sample. For good depth resolution primary ions with low masses and low energy at a grazing impact angle are applied. The impact angle must not be too grazing otherwise the primary ions will be totally reflected. Generally the rule applies to depth resolution that the better the depth resolution the worse the sputter yield.

The problems in quantification are versatile. At the beginning stands the problem that a standard is needed for quantification and that there are only few available. New developments tend to use ion

Analytical Techniques

implantation during measurement in order to derive internal calibration[61]. Further quantification is limited to low concentrations, because the matrix must be equal. The accuracy of quantification is not high either. It amounts about 5% to 10% relative.

When analyzing isolating samples electrostatic charges occur that result from the implantation of primary ions and the emitting of secondary ions and secondary electrons. In conducting samples charging the sample is prevented by a voltage applied to the sample. In order to minimize the effect electron guns are applied to balance the charge. Another possibility is to shift the high voltage applied to the sample so that the sum of the voltages remains constant during measurement. However charging effects influence the ionization probability and cause electrical fields, which affect the path of the secondary ions. So imaging and quantification are impeded.

High vacuum conditions are needed, because impacts of the analyte ions with residual gas molecules or other particles cause a blurred energy distribution of the secondary ions and diminish the transmission of the secondary optics. The mean free path of the ions is a function of the pressure (approx. $\lambda p = 5mmPa$ that means if $p=10^{-4}$ mbar $\lambda=0,5m$) and has to be long enough to avoid collisions between particles. A bigger problem than the mean free path is the high rate of recovering of the sample surface with particles of the residual gas. If the pressure is 10^{-7} mbar it lasts approximately 1s till the surface is recovered with one monolayer of the residual gas (depends on the sticking coefficient). Samples that are destroyed under high vacuum conditions cannot be measured.

Likewise huge samples are not measurable due to the limited size of the sample chamber. Some samples should not be damaged through measurement, because they are either too expensive or of cultural value. Such samples are not suited for SIMS analyses.

The mass spectra of solids are not easily interpreted. There occur a lot of ions such as atom ions of the whole isotopes of an element, molecule ions and cluster ions not only of the analyte but also of the residual gas. So the spectra become quickly confusing.

Analytical Techniques

3.1.4 Quantification

Quantification in SIMS mainly means converting the intensity of the secondary ions into analytical concentrations. The reliance of the secondary ion intensity on the different parameters is given by Equation 2.

$$I_{S(A)} = I_P * S * \alpha_A^{\pm} * i_{S(A)} * \eta_A * c_A \qquad \textbf{Equation 2}$$

$I_{S(A)}$ Secondary ion intensity of the measured isotope of the element A [ions / s]
I_P Primary ion intensity [ions / s]
S Sputter yield [atoms / primary ion]
$\alpha_A^{\pm}$ Positive respectively negative ionization probability of the sputtered atoms
$i_{S(A)}$ Isotope frequency of the measured isotope of the element A
η_A Efficiency of the secondary ion measurement (output of the ion extraction, transmission of the mass spectrometer, efficiency of the detector)
c_A atom concentration of the element A in the sample

For depth profiling an additional calibration of the depth scale is required. The crater depth is measured with a profilometer. It is assumed, that the sputter rate is constant during the measurement.

There are three possibilities in quantification. Two procedures are empirical and one theoretical approach is based on the local thermal equilibrium model.

3.1.4.1 Absolute sensitivity factors

The absolute sensitivity factor (ρ_A) is defined as the change of the secondary ion intensity with the change of the concentration (Equation 3).

Analytical Techniques

$$\rho_A = \frac{dI_{S(A)}}{dc_A} \qquad \text{Equation 3}$$

ρ_A Absolute sensitivity factor
$dI_{S(A)}$ Differential secondary ion intensity of the element A
dc_A Differential concentration of the element A

On account of this equation a calibration curve for the concentration of the element A as a function of the secondary ion intensity is derived. At higher concentrations the curve is not necessarily linear. In addition, constant instrumental parameters are required.

3.1.4.2 Relative sensitivity factors RSF

In order to level out instrumental instabilities and deviations in instrumental adjustment, the concentration of an element A is referred to the concentration of an internal reference element B, in common a main component of the sample. Doing so, the following Equation 4 is derived:

$$\rho_{A/B} = \frac{I_P * S * \alpha_A^\pm * i_{S(A)} * \eta_A}{I_P * S * \alpha_B^\pm * i_{S(B)} * \eta_B} = \frac{I_{S(A)} * c_B}{I_{S(B)} * c_A} \qquad \text{Equation 4}$$

$\rho_{A/B}$ Relative sensitivity factor
I_P Primary ion intensity [ions / s]
S Sputter yield [atoms / primary ion]
$\alpha_A^\pm, \alpha_B^\pm$ Positive respectively negative ionization probability of the sputtered atoms of the element A respectively B
$i_{S(A)}, i_{S(B)}$ Isotope frequency of the measured isotope of the element A respectively elements B.
η_A, η_B Efficiency of the secondary ion measurement (output of the ion extraction, transmission of the mass spectrometer, efficiency of the detector) for the element A respectively B.

Analytical Techniques

$I_{S(A)}$, $I_{S(B)}$ Secondary ion intensity of the measured isotope of the element A respectively B [ions / s]

C_A, C_B.   Atom concentration of the element A respectively B in the sample

For implantation standards the RSF is calculated from the sum of the concentrations in each depth over the whole implantation range, which corresponds to the implantation doses. The RSF is considered constant over the whole concentration range. This leads to the Equation 5:

$$\rho_{A/B} = \frac{\sum_{i=1}^{z} \frac{I_{S(A)(i)}}{I_{S(B)(i)}} * t}{Q_T * z} \qquad \text{Equation 5}$$

z   Number of cycles

$I_{S(A)(i)}$, $I_{S(B)(i)}$ Secondary ion intensity of the measured isotope of the element A respectively B at cycle i [ions / s]

t   Depth [cm]

Q_T Implantation doses [atoms / cm²]

After deriving the RSF from standards and determining the ratio of the secondary ion intensities the concentration of the analyte is calculated with the Equation 6:

$$C_A = \frac{I_{S(A)} * C_B}{\rho_{A/B} * I_{S(B)}} \qquad \text{Equation 6}$$

The RSF is in general, with the exception of high concentrations, constant over some orders of magnitude.

Analytical Techniques

3.1.5 The instrument

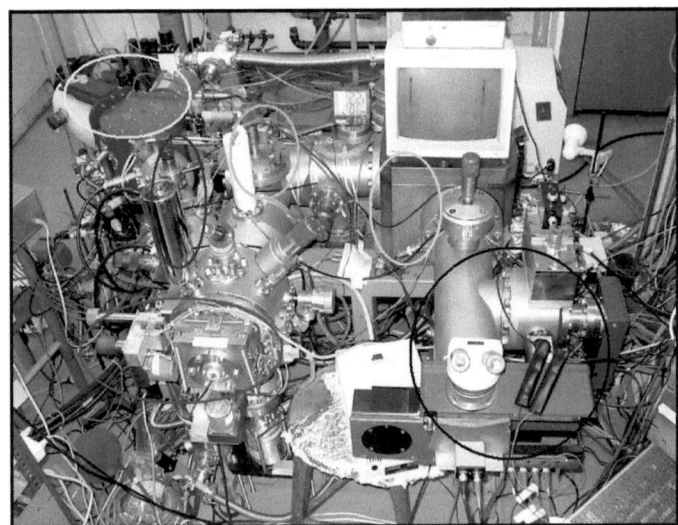

Figure 16: SIMS equipment 'live': primary ion sources and primary ion filter, primary optics, sample chamber, electrostatic analyzer, magnetic analyzer and detection system (from left to right).

3.1.5.1 General

The SIMS instrument consists of several parts. In general it is distinguished between two types of SIMS instruments. On one hand, SIMS instruments as supplementary equipment for instruments such as AES, XPS, LEED, ISS, etc., on the other hand, high sophisticated standalone SIMS instruments called ion microscopes. In this chapter only the latter instruments are discussed. A real photo (Figure 16) and survey over the SIMS instrument, divided into primary and secondary part is shown Figure 17. The main parts are the ion source, the primary ion filter, lens systems which focus and deflect the primary ion beam, the sample chamber with a optical microscope attached, lens systems to extract and focus the secondary ion beam, the electrostatic analyzer, the mass spectrometer, the detection system and, last but not least, an ultra-high- vacuum-pump-system. A

Analytical Techniques

brief description of the different parts of the CAMECA 3f SIMS instrument, which is used in the measurements, follows the Figure 17.

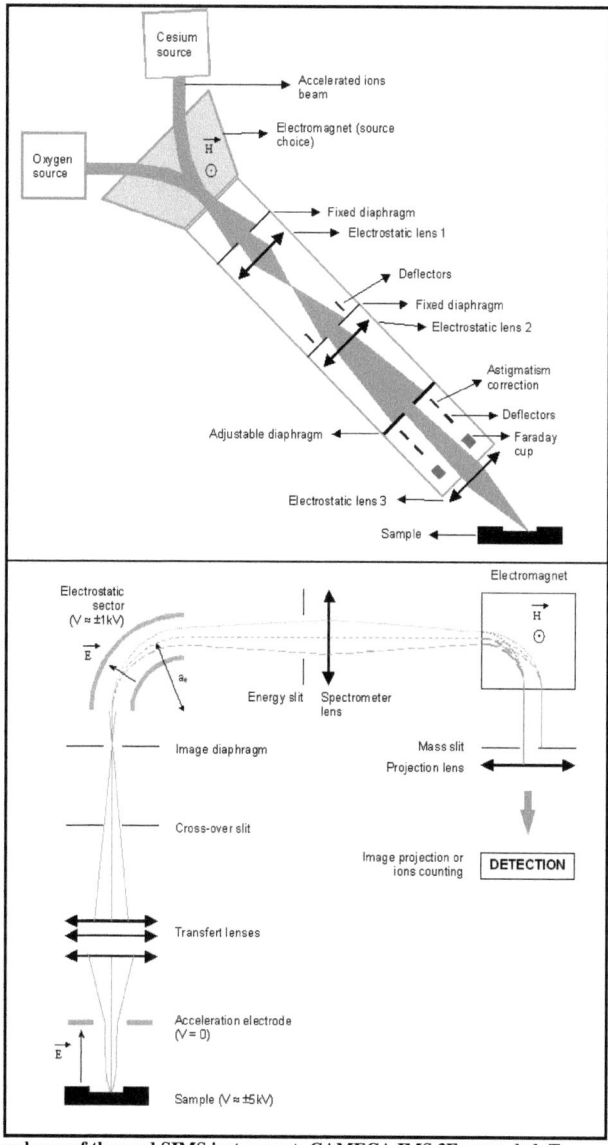

Figure 17: The scheme of the used SIMS instrument; CAMECA IMS 3F upgraded. Top: primary section; bottom: secondary section.

Analytical Techniques

3.1.5.2 Ion sources

The CAMECA IMS 3f SIMS instrument provides two different ion sources, a duoplasmatron (O_2^+, O^-) and a Cs^+ ion source.

In the duoplasmatron (Figure 18) an oxygen plasma is produced by a discharge processes between a cathode and an anode. Positively and negatively charged ions as well as electrons emerge from the plasma. The ions are extracted by an extraction electrode. The magnetic field drives the electrons on spiral paths, which enhances the affectivity of the ionization.

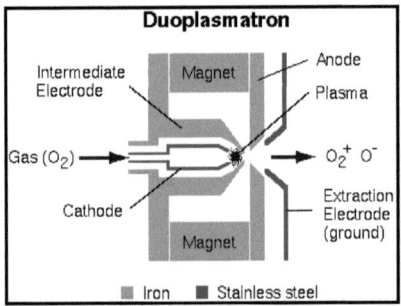

Figure 18: The scheme of duoplasmatron ion source

The second ion source is the Cs^+ liquid metal ion source (Figure 19). Cs is heated and vaporized in the reservoir. The Cs vapour diffuses into a porous tungsten plug, which is warmed to 1100 °C. Cs is ionized there by thermal surface ionization. The positively charged ions are focused and extracted by electrodes.

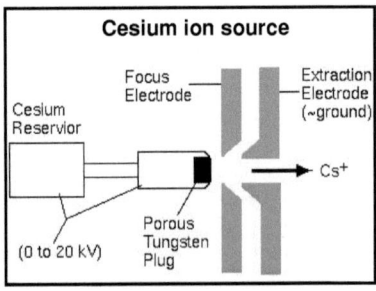

Figure 19: The scheme of Caesium ion source

Analytical Techniques

3.1.5.3 Primary ion optics

The primary ion optics (Figure 20) consist of a mass filter, which cares for the separation of contaminations and allows isotope-pure measurements, sometimes used in O_2^+ measurements, electrostatic lens and aperture systems, to focus and deflect the primary ion beam onto the sample.

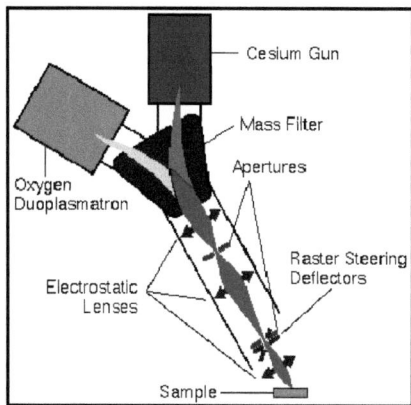

Figure 20: The scheme of primary ion column

Electrostatic lenses consist of two end plates and one active element (Figure 21):
  The end plates operate at ground potential.
  The centre element operates at high voltage.

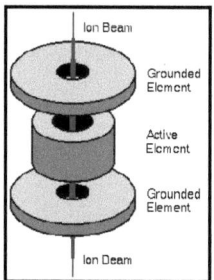

Figure 21: Schematic construction of electrostatic lenses

Analytical Techniques

3.1.5.4 Secondary ion extraction system

The extraction system (Figure 22) extracts the secondary ions and focuses them onto the entrance slit of the mass spectrometer. The main parts are electrostatic lenses and apertures. In common, the extraction voltage is provided by a high voltage applied to the sample whereas the extraction lens is kept on ground potential.

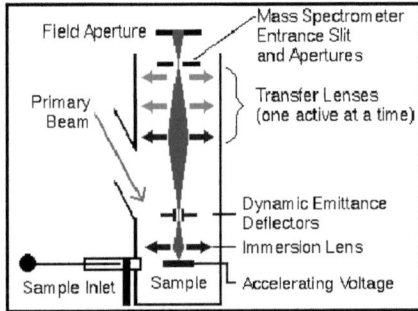

Figure 22: The secondary ion extraction system.

3.1.5.5 Double focusing mass spectrometer

The mass spectrometer is divided into two parts: the energy analyzer (Figure 23), which is next to the secondary ion extraction system, and the mass analyzer, where the ions are separated by their mass to charge ratio. In the energy analyzer a homogeneous electrical field is applied. Thus, an electrical force depending on the charge of the ions forces the ions with different speeds, that means different kinetic energies, and the same charge onto different paths. Only ions with the same speed to charge ratio can pass the energy slit.

Analytical Techniques

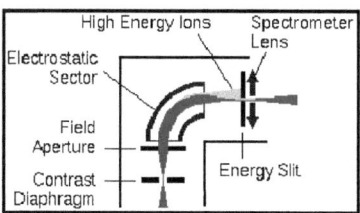

Figure 23: The energy analyser

In the mass analyzer (Figure 24) a magnetic field perpendicular to the trajectories of the ions is applied. The ions are separated by the Lorentz force according to their mass to charge ratio.

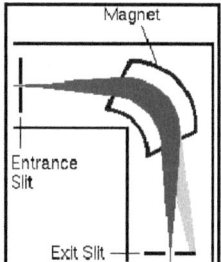

Figure 24: The magnet mass analyser.

3.1.5.6 Detection system

The detection system (Figure 25) consists of apertures, lenses, an electrostatic sector, which provides the capability to switch between the different detectors, such as channel plate, electron multiplier and the faraday cup.

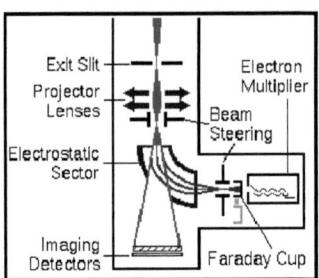

Figure 25: The different detectors

The channel plate (Figure 26) is built of about 200 parallel tubes, which collect the ions and multiply the signal. So a lateral resolution is achieved. The principle is that of an electron multiplier. Ions impact the inside wall of a tube and knock out electrons. The electrons are accelerated by a difference in potential, colliding again with the wall and so the number of electrons, multiplied. The signal will be enhanced.

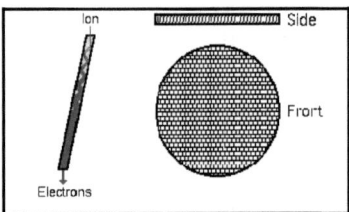

Figure 26: The channel plate.

Electrons impinging the electron multiplier (Figure 27) knock out electrons of a dynode. A potential difference is applied between successive dynodes due to that electrons were accelerated and hit the following dynode with a higher energy. Therefore, more electrons were emitted from one dynode to the next. So the signal is enhanced.

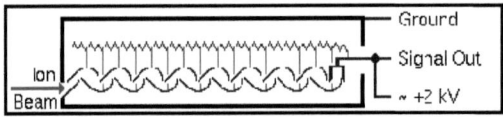

Figure 27: The electron multiplier.

Analytical Techniques

The Faraday cup (Figure 28) extends the dynamic range for the detection of secondary ions. When the intensity is too high for the electron multiplier the ions are deflected into the Faraday cup. The signal is direct proportional to the number of secondary ions.

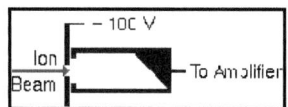

Figure 28: The Faraday cup.

3.1.5.7 The modified Cameca IMS 3f

The SIMS instrument used throughout these investigations was an enhanced CAMECA IMS 3f .The modification on one hand mainly relates to the primary ion column, an additional primary magnet was installed to use a CAMECA Cs^+ fine focus ion source as well as O_2^+ duoplasmatron. On the other hand, the original beam deflection unit was replaced by a digital scan generator (DGS)[62] communicating with the controlling workstation via a specially designed parallel interface unit. In conjunction with the newly developed SIMS software package the interface unit also features the capability to control the secondary magnet, the ion projecting and focusing lens system and performs the readout of the electron multiplier (EM) and the Faraday cup.

3.1.5.7.1 The new interface

The new interface unit allows full access to all electronically controllable parts of the instrument. The unit also controls the secondary magnet, the ion projecting, the focusing lens system and it performs the readout of the electron multiplier and the Faraday cup. The output of the electron multiplier is transferred via a fiber optical link to the controlling workstation. It would also be possible to control a stepper motor to move the sample automatically to perform line scans over very large distances, which would not be possible only by deflecting the beam.

49

Analytical Techniques

3.1.5.7.2 The new scanning generator

The original beam deflection unit was replaced by a digital scan generator (DSG) communicating with the controlling workstation via a specially designed parallel interface. It is capable of doing the automatically fast scanning that is necessary for stigmatic SIMS over areas up to 1000 µm as well as the computer-controlled digital step scanning that is necessary for the scanning mode.

3.1.5.7.3 The new ion source

The original Cs-metal-ion source was replaced by a combination of a CAMECA Cs^+ fine focus ion source and an O_2^+ duoplasmatron with a primary magnet to select between the two available sources. Furthermore, the application of the magnet guarantees the purity of the primary ion species. The Cs^+ ion gun is capable of supplying up to about 0.3 µA primary current, the O_2^+ duoplasmatron up to 6 µA.

3.1.5.7.4 The improved vacuum system

Together with the implementation of the new ion source a completely new primary column with an improved vacuum system has been built to guarantee high vacuum conditions in the sample chamber ($< 10^{-9}$ mbar). Thus, low residual gas interferences and slow coverage of the sample surface with the residual gas are achieved. Such measurement conditions are a stringent demand for trace element analyses and scanning mode operation. The original attached turbo molecular pump has been replaced by a cryogenic pump. In order to maintain low pressure in the primary column during O_2 measurements, differential pumping is demanded, because the pressure in the duoplasmatron is about 10^{-5} to 10^{-4} mbar. Therefore, two turbo molecular pumps, one directly after the duoplasmatron and the other in the middle of the primary ion column, and an additional cryogenic pump have been attached.

Analytical Techniques

3.2 Rutherford Backscattering Spectrometry – RBS[58]

3.2.1 Principle of RBS

Rutherford backscattering (RBS) is based on collision of monoenergetic ions, usually H^+ or He^+ with the stationary sample atoms. The energy of the primary ions is in the range from 0.5 to 3 MeV. The impacting projectiles transfer their energy to the stationary sample atoms and are then subsequently backscattered. The number and the energy of the backscattered projectiles is then measured and allows the determination of the atomic mass and elemental concentrations depending on depth where the impacting projectiles have been backscattered. The principle is schematic demonstrated in the Figure 29[63].

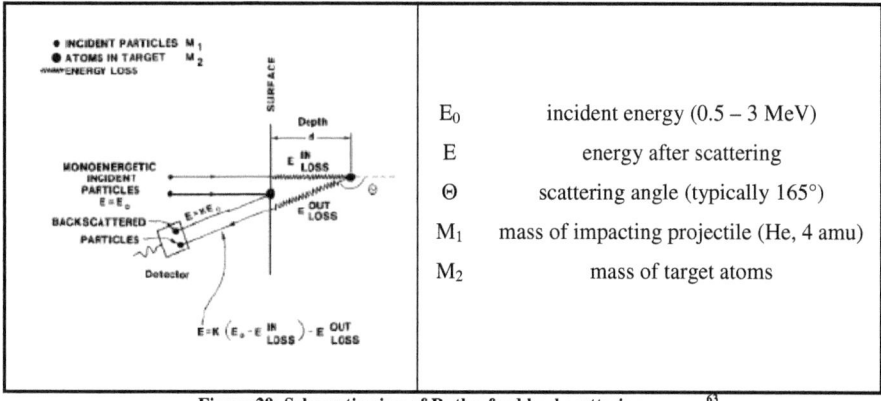

Figure 29: Schematic view of Rutherford backscattering process[63].

3.2.1.1 Kinematics, the cross section and the stopping power

When a primary ion projectile is impacting the sample, the strong Coulombic forces are responsible for the repulsion. At the present energies (0.5 – 3 MeV) there is no reaction between the primary ion nuclei and the target atom nuclei thus interaction can be treated as classical elastic collision.

Due to the collision, the primary ions lose a part of their energy. The energy loss depends on mass ration of the participating nuclei and with the knowledge of the beam, target and detector geometry it is possible to determine the sample composition. If He^+ are impacting a sample consisting of silicon and germanium they will lost more of their energy at the atoms with lower mass (silicon).

If primary ions are scattered directly on the surface the only loss of energy is the momentum transfer to the target atoms. The kinematic factor (K) defines the energy ratio of the projectiles after and before scattering (Equation 6).

$$K = \frac{E_{SCATTERED}}{E_{INCIDENT}} = \left[\frac{\left(1-\left(\frac{M_1 \sin\theta}{M_2}\right)^2\right)^{\frac{1}{2}} + \frac{M_1 \cos\theta}{M_2}}{1+\frac{M_1}{M_2}} \right]^2 \quad \text{Equation 7}$$

E   Ion energy
M_1 Mass of incident ion
M_2 Mass of target atom
θ   Scattering angle

With increasing mass of target atoms a less momentum is transferred from the impacting projectile and the energy of the backscattered projectiles asymptotically approaches to the incident projectile energy. The consequence is that RBS is more useful tool for distinguishing between two light elements then between two heavy elements.

The amount of the backscattered projectiles from a target atom into certain solid angle for a certain number of incident projectiles is depending on the differential cross section. The scattering cross section is proportional to the square of the atomic number of the target atom

Besides their loss of energy at the collision the primary ions lose their energy also on the way to the target atom, within the sample, but also on the way out of the sample. This loss of energy depends on the materials stopping power. The penetrating projectiles at first are interacting with the electrons in the surrounding material (electronic stopping) and second the energy dissipates due to glancing collisions with the nuclei of target atoms (nuclear stopping). Theoretical predications of stopping power are very complicated and sometimes very inaccurate. Therefore, empirical stopping powers are often used in RBS calculations, but nevertheless the simulation software as SIMNRA[64] is also often used to calculate RBS spectra.

3.2.2 Instrumentation and applications

RBS instrument consists of four main components:

- *Primary ion source (He ions)*

If single-ended accelerators is used the ion source is on the floating at high voltage. Electrical isolation of the megavolt potentials is needed and is achieved by housing the terminal in a tank filled with an insulating gas, usually SF_6. One disadvantage of locating the ion source within the tank is that it is difficult to change or replenish the source. If tandem accelerator is used a positive terminal is located in the centre of the device. Negatively charged particles are injected into the accelerator and attracted to the terminal where a stripper element removes two or more electrons from each particle. The positive terminal repels the resulting positive ion back toward ground.

- *Primary ion accelerator*

The widely used RBS instruments use a Van de Graaff electrostatic accelerator either single ended or double ended (tandem)[65]. The accelerator produces the ions in several charge states and even a multiplicity of species, which subsequently pass field of analyzing system in order to extract the beam suitable for material analysis. Afterwards mass and charged selected beam enters the UHV vacuum system and is then oriented towards the target. 1 mm² is a typical beam size on the target, but using suitable lens system it is possible to achieve smaller spot sizes.

Analytical Techniques

- *Sample chamber*

The sample chamber consists of a stage area (typically mounted on a five axis goniometer), one or more detectors, beam line entrance and the vacuum system and is schematically shown in the Figure 30.

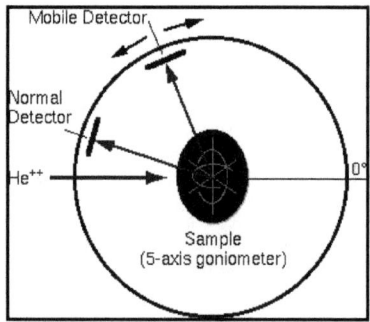

Figure 30: Schematic overview of the incoming He^{2+} beam into the sample chamber with normal and mobile detector.

- *Detecting system*

The detection of the backscattered ions is usually performed with a solid state detector, mostly used passivated implanted planar silicon (PIPS) detector.

With sufficiently good depth resolution of approximately 10 nm (mention: only using low energy RBS or using grazing incidence angle) and the advantage of standard – free quantification RBS is a widely used quantitative technique to analyse the composition and thickness as well as for depth profiling of thin film or solid samples in the surface near region. RBS is often used as quantitative feature for other techniques e.g. SIMS. Generally the measurements are simple, due to usage of relatively small accelerator, and quantitative but subsequently interpretation and evaluation requires computer simulations by means of generally iteration method. The commonly used software are as mentioned above SIMNRA[64] and RUMP[66]. Nevertheless there are also some drawbacks of the technique e.g. intermix of mass and depth information, no hydrogen analysis, light elements (B, C, N, O, F) are only badly detectable inside the heavy matrix and the sensitivity for heavy elements is better then for light ones. Table 1 shows the summarised figures of merit of RBS.

Analytical Techniques

Table 1: RBS – figures of merit

Primary probe	Elemental range	Type of inf.	Depth of inf.	Lateral resol.	Sensitivity [at%]	Insulator analysis	Destructive	UHV
H^+, He^+	high Z on low Z	Elemental, depth distr.	1ML – 2 µm	5 µm – 1 mm	1e-6	Yes	Yes	Yes

3.3 Scanning and Transmission Electron Microscopy – SEM / TEM

3.3.1 General

The interaction of high energy electrons (typically 20 keV) with a certain kind of sample can result in different phenomena. The electrons can transmit the sample losing a part of their energy, the electrons can be reflected on the sample surface or inside the sample and the electrons can be absorbed within the sample. Further on the sample can emit electrons, photons and charged ions. The interaction of an electron beam with the atoms of an analyte is shown in Figure 31.

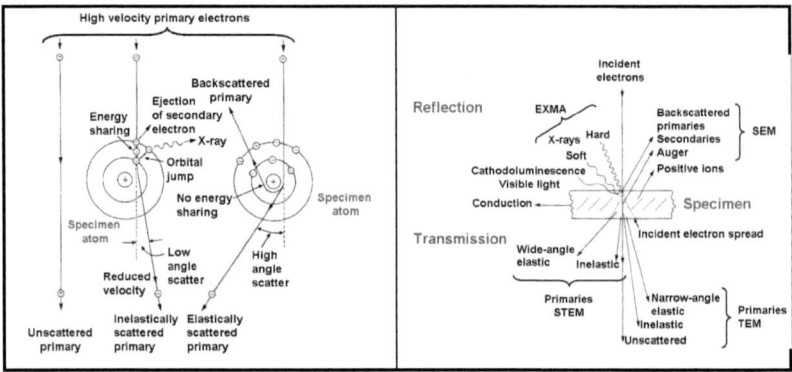

Figure 31: The principal modes of interaction of high energy electrons with sample atoms[67].

The electron microscopy is widely used term for a couple of surface analytical techniques based on the electron excitation of the sample. In general the primary electron beam can transmit the sample (Transmission Electron Microscopy – TEM) or it can be reflected from the sample (backscattered electrons – typically termed SEM/backscattered electron mode) or even secondary electrons can be excited (secondary electrons – typically termed SEM/secondary electrons mode). The primary electron beam can be focussed down to a diameter of 1 – 100 nm which leads to very high lateral resolution and due to a higher cross section, compared with photons, here higher signals are obtained. The above mentioned two modes of electron microscopy, TEM and SEM will be shortly described in the following chapter.

Analytical Techniques

3.3.2 TEM instrumentation and applications[68]

Typically TEM instruments consist of illuminating part (electron source, condenser and objective lenses), sample stage, imaging system (intermediate and projector lenses, fluorescent screen with camera) and a vacuum system.

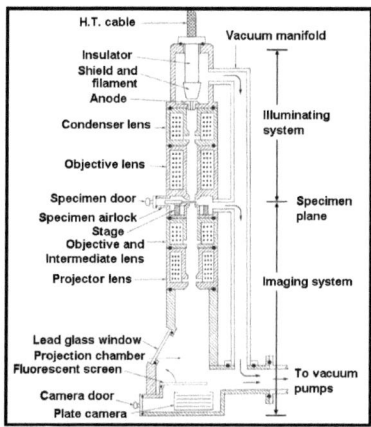

Figure 32: Schematic overview of a TEM instrument.

Nowadays the field emission guns (a sharply pointed tungsten tip) replace the widely used thermionic electron guns (tungsten, LaB_6, Ce_6). Field emission guns use strong electrical fields (up to 10^9 V/m) to produce and to extract electrons from metal filament. The operating temperature is lower (room temperature) then in thermionic guns (2500°C), but specifically a good ultra high vacuum is required. The brightness of the field emission guns is 5 orders of magnitude higher than of thermionic electron guns. The thermionic electron guns are more robust and cheaper and no ultra high vacuum is required. Here the electrons are emitted from a heated filament and subsequently accelerated towards an anode. Further electron guns are thermal field emitters (the pure field emission is enhanced by addition of some thermal energy) and a Schottky emitter (thermal field emitter with a doped surface in case of reduction of the work function).

Compared to optical microscopy the magnetic lenses replace glass lenses. Generally the magnetic field can be generated if current is passing through of a set of windings. The resulting magnetic field acts as a thin convex lens. In the illuminating part of the TEM instrument two (double condenser lens) or more of such lenses completed with a condenser aperture (control the intensity and the area of the illumination) are used to control the spot size and the resulting beam convergence.

The sample stage is commonly situated in the objective lens area and can be moved in x, y, as well as in z direction and it also can be rotated and tilted.

After going through the sample the electrons have to pass the imaging part of the TEM instrument. The objective lens forms an inverted initial image, which is subsequently magnified and additionally the diffraction pattern is formed in the back focal plane of the objective lens. Here the objective aperture is placed to enhance the contrast of the final image but more to select the electrons which will be used to form the image. The magnification in TEM can be varied from hundred to hundred thousands of times. This can be done by the usage and variation of strength of the projector and intermediate lenses. In particular the intermediate lens is responsible for the magnification of the initial image.

TEM is a unique tool for characterisation of crystal- and microstructure of materials simultaneously by diffraction and imaging techniques. The typical electron energies are from 120 up to 500 keV. The energy of the accelerated electrons in TEM is responsible for their penetration inside and through a specific sample, thus the sample thickness usually does not exceed 2µm. For the preparation of samples ion beam milling, cryofixation, embedding, sectioning, staining and as well as wedge polishing techniques are used. The typical application area is in the material science and biological science. TEM instruments provide detailed images of investigated samples as well as diffraction patterns and TEM is a complementary tool to conventional crystallographic methods.

Nevertheless there are some disadvantages in particularly expensive to buy and to maintain. TEM instruments are sensitive to vibration and external magnetic fields, thus it needs a specific place to handle with it. Due to a vacuum requirement it is near impossible to examine living material.

3.3.3 SEM instrumentation and applications[69]

Scanning electron microscopy (SEM) is a related technique to TEM. The electron source and magnetic lenses are comparable to those of the TEM instruments as described above. The mean difference is that a fine focussed electron beam is scanned over the sample and no electrons are transmitting the sample. The information obtained from each scanned point is used for the characterisation of material or for the image generation. The schematic overview of a SEM instrument is given in the Figure 33.

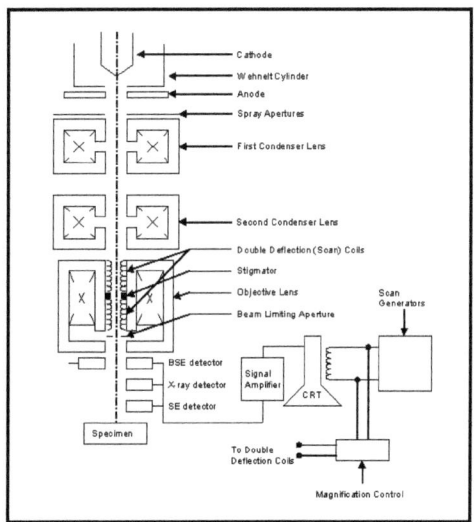

Figure 33: Schematic view of the Scanning Electron Microscope[70].

The typical electron energies here are in the range from 20 – 100 keV and the electron beam is focused to a spot size of approximately 5 nm by means of two condenser lenses. Passing the objective lenses the beam is scanned over the sample by means of scanning coils.

When the electron beam interact with the sample surface the electrons are inelastically scattered by the sample atoms ant the beam is spread consequently within the sample whereby the interaction volume is up to 5 µm. Hereby the primary electrons can be scattered and detected after these have leaved the sample. Backscattered electrons can be used to compose the topographical image of the sample but also to distinguish between different chemical compositions of the sample. The backscattered electrons have nearly the same energy as the primary electrons and these are typically scattered in the deeper regions of the sample. The backscattering of the primary electrons is a function of the atomic number of the sample atoms and therefore the chemical composition of collected backscattered electrons can be obtained.

Further on the emission of secondary electrons from the sample, after this one is stroked with primary electrons, is also possible. The secondary electrons originate from the first 20 nm of the sample surface and have lower energy (< 50 eV) than the backscattered electrons and most images made by SEM monitors this secondary electrons. The brightness of the signal is a function of the surface area that is exposed to the primary beam, thus some higher situated edges on the sample appear brighter.

Besides the backscattered electrons and secondary electrons, the interaction of electrons with the sample can also as showed in Figure 31 result in phonon excitation, cathodoluminescence, continuum X – ray radiation, Auger electron emission but also in characteristic X – ray radiation. This one is used to detect the chemical composition of the sample without any other reference materials. The specific X – rays can be by wave length dispersive detector (WDS) or by energy dispersive X – ray detector (EDS).

The SEM instruments provide superior resolution when compared with an optical microscope and provides very detailed images which have 3D structures. The spatial resolution of the SEM instruments depends primarily on the selected electron beam spot size. The imaging down to the atomic scale as it is possible by TEM is here, due to large interaction volume, not available but the resolution of 1 to 5 nm is possible. The sample preparation is quite simple compared to TEM as long the samples are conductive. The SEM technique is, belong other research areas, the method of choice for the analysis of the fractured surfaces.

3.4 Auger Electron Spectroscopy

3.4.1 Principle of AES [58]

Auger Electron Spectroscopy (AES) is one of the most used surface analytical techniques. It provides the chemical composition of thin films, grain boundary segregations, surfaces of solids and thin coverages of surfaces. Under electron bombardment with a focused electron beam (energy range 2 – 10 keV, beam diameter 50 – 500 nm, vacuum required < 10^{-9} mbar) at first electron from the core level of analysed samples will be removed. This induces a fall of the electron from the higher level into the vacancy of the inner shell. The energy excess caused by the two previous processes has to be released immediately. The first possibility is the emission of the Auger electron, which have characteristic kinetic energies depending on the emitting chemical element. These electrons are then collected and measured. Figure 34 shows a principle of AES The second possibility and the competing process to the Auger electron emission is the emission of an X – ray photon (Figure 34). For shallow core levels the probability of the Auger process is higher then of the X – ray emission process.

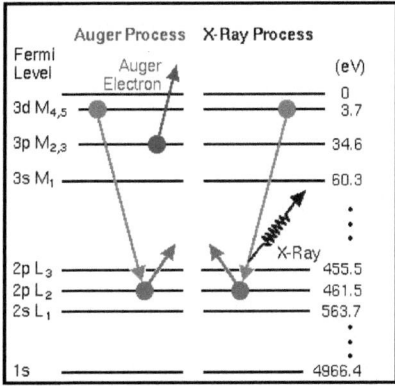

Figure 34: The scheme of Auger process and X – ray emission.

The mean projection range of the electron beam striking the sample surface is at about 1 – 5 µm. X – ray photons are emitted from the whole volume, but Auger signals are very sensitive and come only from the first 2 – 6 monolayers (Figure 35).

Analytical Techniques

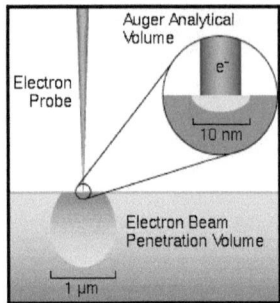

Figure 35: Auger analytical volume compared with the electron beam penetration volume.

3.4.2 Instrumentation and Applications

All electron spectroscopic techniques require vacuum conditions with a pressure level of 10^{-10} mbar (ultra high vacuum, UHV) for their operation, otherwise there could be interactions of the electrons on the way to the sample, interactions of the electrons with another atoms and molecules which are present inside the instrument (residual gases), interactions of the analyte with those atoms ore molecules (surface coverage and contamination) which all lead to the false interpretation of obtained results.

The usually used primary electron energies are in the range of 3 – 10 keV and the electron source is either tungsten filament cathode or LaB_6 cathode which provides higher current densities, which allows narrower electron beams useful for analysing smaller features. Field emission electron guns provide the brightest beams and are most advantageous achieving minimum spot sizes. Afterwards the electron beam is focused electromagnetically.

An electron – energy – analyser measures the number of ejected electrons depending on their energy. The cylindrical mirror analyzer (CMA) is also located in one high vacuum chamber (the reason are the same as described above) and must be isolated from the stray magnetic fields. The scheme of the CMA is shown in the Figure 36.

Analytical Techniques

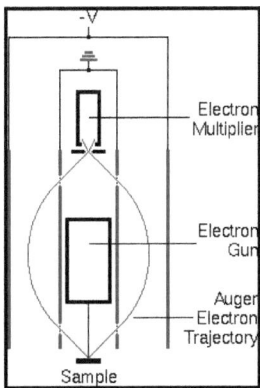

Figure 36: The scheme of cylindrical mirror analyzer.

Due to the characteristic kinetic energies of emitted Auger electrons, AES is one of the essential techniques for the detection of chemical composition of the analyte for the first monolayers. The analytes can be different e. g. surfaces, thin films but also interfaces. If used in depth profile (sputtering with argon ions) mode the technique is then destructive. The main disadvantage is the analysis limited only for the conducting materials. Due to the principles of the technique (3 electrons are needed) it is not possible to detect hydrogen or helium. One enhancement of the AES method is Scanning Auger Microscopy (SAM). Here the electron beam is fine focussed and deflected over the sample, whereby element specific images (maps) are obtained. Nevertheless the line scans and small feature analysis (very small spot size and very large magnification) can also be performed. Figure 37 is showing a schematic diagram of a common used Auger electron spectroscopy instrument and the figures of merit are summarised in the Table 2.

Table 2: AES – figures of merit

Primary probe	Elemental range	Type of inf.	Depth of inf.	Lateral resol.	Sensitivity [at%]	Insulator analysis	Destructive	UHV
e⁻, 3-10 keV	All except H and He	Elemental (chemical)	2 - 6 ML	20 nm	0.2	No	Yes	Yes

Analytical Techniques

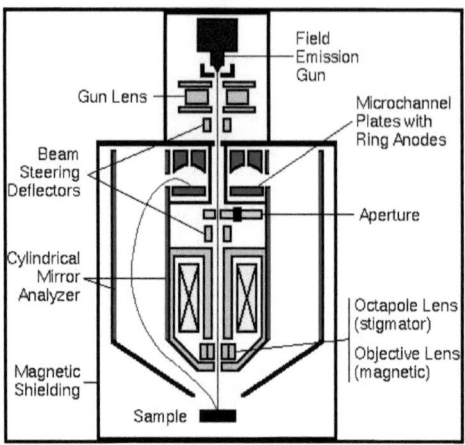

Figure 37: Schematic diagram of a common Auger electron spectroscopy instrument.

4 Sample preparation

4.1 PM samples

The investigated PM samples in this work were produced out of three different water atomised steel powders: ASC 100.29 (pure iron powder), Astaloy CrM (containing 3 wt% Cr and 0.5 wt% Mo) and Astaloy CrL (1.5 wt% Cr and 0.2 wt% Mo), all produced by Höganäs AB, Sweden[5, 6]. Except the reference samples, all the other samples contain either boron (ferroboron, Fe18B, particle size <40µm) or phosphorus (ferrophosphorus, Fe_3P, particle size <40 µm) acting as sintering activator and depending on the investigating topic in some cases also carbon (UF 4 nature graphite).

The base powders, used sintering activator, carbon and pressing lubricant (HWC) were blended for 60 min in a tumble mixer. The compaction of the green compacts (55mm x 10mm x 10mm) was made by uniaxial pressing at 600 MPa. After de-waxing at 600°C for 30 min the green compacts were then sintered for 60 min in a pusher furnace in one certain sintering atmosphere. The green densities were calculated from the measured specimen bars dimensions (length, width, height) and mass. Sintering density measurements were made after resin impregnation by means of water displacement method. Table 3 gives the overview of the sample production and sample composition of the investigated samples.

Table 3: Overview of the sample preparation and composition fort he PM samples ([+]... concentration steps: 0, 0.03, 0.06, 0.1, 0.15, 0.2, 0.3, 0.6; [++]... concentration steps: 0, 0.15, 0.3, 0.45, 0.6, 0.8, 1)

sintering activator	base powders	sintering temperatures	sintering atmosphere	use of carbon	sintering time
Boron (0 – 0.6 wt%)[+]	ASC 100.29 Astaloy CrL Astaloy CrM	1200°C 1300°C	hydrogen, nitrogen, argon	yes and no	60 min
Phosphorus (0 – 1 wt%)[++]	ASC 100.29 Astaloy CrL Astaloy CrM	1120°C 1250°C	hydrogen, argon	yes and no	5, 10, 15, 20, 30, 60 min

4.2 Tribology

The substrates for the investigated samples were standard aerospace bearing steel AMS 6491 and high strength stainless steel grade AMS 5898. The chemical composition of these two materials is presented in Table 4. The samples were hardened due to austenitizing and subsequently quenching in oil. Afterwards deep frozen step followed and finally the samples were annealed as presented in Table 4. This treatment ensures a constant hardness of 60 HRC and sufficient corrosion resistance of AMS 5898.

Table 4: Chemical sample composition and temperature treatment during the sample production. (Austen.temp. – Austenitisation temperature; DF.temp. – Deep Freeze temperature; Anneal.temp. – Annealing temperature).

Sample	C [wt%]	N [wt%]	Cr [wt%]	Mo [wt%]	V [wt%]	Austen. temp	DF. temp	Anneal. temp.
AMS 6491	0.82	-	4.1	4.2	1	1100°C	-70°C	585°C 3 x 2h
AMS 5898	0.3	0.4	15.2	1	-	1030°C	-70°C	185°C 2 x 2h

In order to study the wear behaviour of the steel grades ball on disc (BOD) tests were performed. A heat treated discs were grinded and polished (1µm diamond suspension) and completely immersed under well defined amount (7 mL => 5mm oil height over the sample) of Mobil Jet II oil[16]. BOD tests were applied at room temperature and at 150°C. As sliding contact partner served Ø 6 mm balls made of AMS 6491. The applied loads were 2 and 15 N. To compare the reaction layer formation the BOD test with tests under rolling contact conditions, rolling contact fatigue (RCF) tests were performed. The rod samples were turned with a size of Ø 9.93 mm and subsequently heat treated and grinded down to Ø 9.5 mm. As counterpart for the rolling contact fatigue tests AMS 6491 steel balls were used. The RCF tests were stopped after 10^6 loads. A constant dripping lubrication was supplied (9 – 11 drops per minute), feeding Mobil Jet II oil. To obtain the information about the surface and in depth composition SEM and SIMS measurements were performed.

Sample preparation

4.3 Ion implantation – samples

This part contains two different research areas. In the first part the difference of impurity gettering after phosphorus (P$^+$) and silicon (Si$^+$) ion implantation was investigated.
Si$^+$ (^{28}Si) and P$^+$ (^{31}P) were implanted into (100) p- type Czochralsky (CZ) silicon. The implantation energy was 3.5 MeV. TRIM predicted mean projected ion ranges were 2.43 µm for P$^+$ respectively 2.46 µm for Si$^+$. The implanting angle was 7° and the implantation dose 5*10^{15} at/cm² except the sample 5 where the implantation dose was one order of magnitude lower (5*10^{15} at/cm²). After the implantation procedure the samples were annealed at 900°C for 5 respectively for 20 min. In order to investigate the residual damage copper was implanted subsequently from the backside (20 keV, 3*10^{13} at/cm²) of the samples. Subsequently heat treatment for 3 min at 700°C allows the completely copper distribution inside the samples (sample thickness 0.5 mm; diffusivity D (Cu) = 4.3*10^{-5} cm²/s @ 700°C)[71], which afterwards was investigated by means of SIMS. Table 5 gives a summary of sample preparation with all proceedings conditions.

Table 5: Sample preparation for investigation on gettering effects at R$_P$, R$_{P/2}$ and trans R$_P$.

Sample	matter	implantation dose [at/cm²]		energy [MeV]	process steps	
		P$^+$	Si$^+$		annealing	dose [at/cm²] ^{63}Cu$^+$
1	p-type Si	-	5*10^{15}	3.5	900°C/20min	3*10^{13}
2	p-type Si	-	5*10^{15}	3.5	900°C/5min	3*10^{13}
3	p-type Si	5*10^{15}	-	3.5	900°C/20min	3*10^{13}
4	p-type Si	5*10^{15}	-	3.5	900°C/5min	3*10^{13}
5	p-type Si	5*10^{14}	-	3.5	900°C/5min	3*10^{13}

In the second part oxygen was implanted into (100) p – type CZ silicon. The implantation energy was 200 keV and the implantation angle was 22°. The ion fluence was 1*10^{17} at/cm². Except the reference sample all samples were subsequently annealed at 850°C for 5h in the argon atmosphere. In order to study the defect engineering for ion beam synthesis of SOI structures samples with pre implanted helium (4*10^{16} at/cm² @ 45 keV) but also with pre implanted helium and additionally simultaneous implantation of oxygen and silicon (3.5*10^{15} at/cm² @ 1 MeV) was investigated. One sample with pre implanted helium (8*10^{15} at/cm² @ 45 keV) and subsequently implanted oxygen

Sample preparation

was further on oxidised in order to evaluate the oxygen amount which can be introduced inside the samples by means of diffusion from the surface. Table 6 presents the sample preparation for the second part of ion implantation and defect engineering samples.

Table 6: Preparation and additionally after treatment for samples made for investigations on defect engineering for ion beam synthesis of SOI structures.

Sample	matter	implantation species			additional treatment
		oxygen [at/cm²]	helium [at/cm²]	silicon [at/cm²]	
A	p-type Si	$1*10^{17}$ 200 keV	-	-	-
B	p-type Si	$1*10^{17}$ 200 keV	-	-	annealed (850°C, 5h, Ar)
C	p-type Si	$1*10^{17}$ 200 keV	$4*10^{16}$ 45 keV	$3.5*10^{15}$ 1 MeV	annealed (850°C, 5h, Ar)
D	p-type Si	$1*10^{17}$ 200 keV	$8*10^{15}$ 45 keV	-	annealed (850°C, 5h, Ar) oxidised (7min, in wet atmosphere, 1150°C)

4.4 $Si_{1-x}Ge_x$ heterostructures samples

The $Si_{1-x}Ge_x$ samples were produced by molecular beam epitaxy (MBE). The samples were grown in an ATOMICA MBE device supported with an electron beam evaporator for silicon and germanium and an effusion cell for antimony doping[72, 73, 74]. Typical growth rates were in the range of 2.5 Å/s for SiGe –, doping – and cap – layer and 0.5 Å/s for the active quantum well layer. The background pressure during epitaxy was kept at about 10^{-10} Torr and the epitaxy temperature was 750°C (550°C for quantum well). The used substrate, p – type Si, was chemically pre – cleaned by etching and subsequently grown protective oxide was thermal (900°C) removed prior to the epitaxy growth. The doping element Sb was introduced by secondary implantation keeping the implantation

Sample preparation

energy below 350 eV to avoid the ion implantation damage. The composition of the investigated samples is described in the Table 7 and Figure 38.

Table 7: Summary of analysed SiGe heterostructure samples. All concentration values are expressed as atom percentage (at%). Note that the layer written in italic font is the doping layer, whereby the nominal concentration of the doping element antimony was 0.25 at%.

layer designation	layer thickness [nm]	Sample A		Sample B		Sample C	
		Si	Ge	Si	Ge	Si	Ge
cap layer	10	100	0	100	0	100	0
SiGe	25	75	25	75	25	75	25
doping layer	*15*	*75*	*25*	*75*	*25*	*75*	*25*
SiGe	10	75	25	75	25	75	25
quantum well	12	100	0	95	5	90	10
SiGe	500	75	25	75	25	75	25

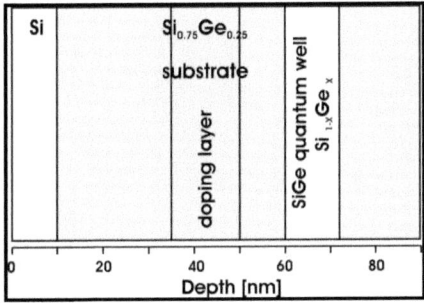

Figure 38: Overview of the layer structure of the analysed samples containing $Si_{1-x}Ge_x$ heterostructures.

The reference sample, here used as the response function of the SIMS instrument, was Ge – δ – layer. The reference sample was also produced by means of MBE in similar way as described above. The substrate was also p – type Si and then one Ge layer was grown. Subsequently 50 nm Si was grown which are essential to reach the sputter equilibrium until the δ – layer is reached during the sputtering process. The growth rates were here 1 Å/s for Ge and 2.5 Å/s for Si.

Sample preparation

5 Results and discussion

The results which will be shown here are the extraction of all obtained results. A reference to the already published papers will be made inside the each handled chapter.

5.1 Powder metallurgy

Analysing the powder metallurgical samples, at first the time (time is a very important factor for the phosphorus diffusion into iron grains) was investigated, which was at least needed for sintering the green parts. It could be shown that the after 5 or 10 min inside the sintering zone the originally used Fe_3P particles are still covered with oxygen, the formation of a liquid phase and subsequent diffusion are inhibited, and there is almost no reaction with the matrix element iron. With increasing sintering time the oxide layer of Fe_3P particles is almost completely destroyed and phosphorus interacts more and more with the iron matrix. After 20 minutes sintering there are still areas observable with lower phosphorus content and finally after 30 minutes (a completely loss of covering oxide layer) there is almost homogeneous distribution of phosphorus observable (Figure 39). For ensuring the equal sintering conditions all samples investigated throughout this work were afterwards sintered for 60 min.

Results and discussion

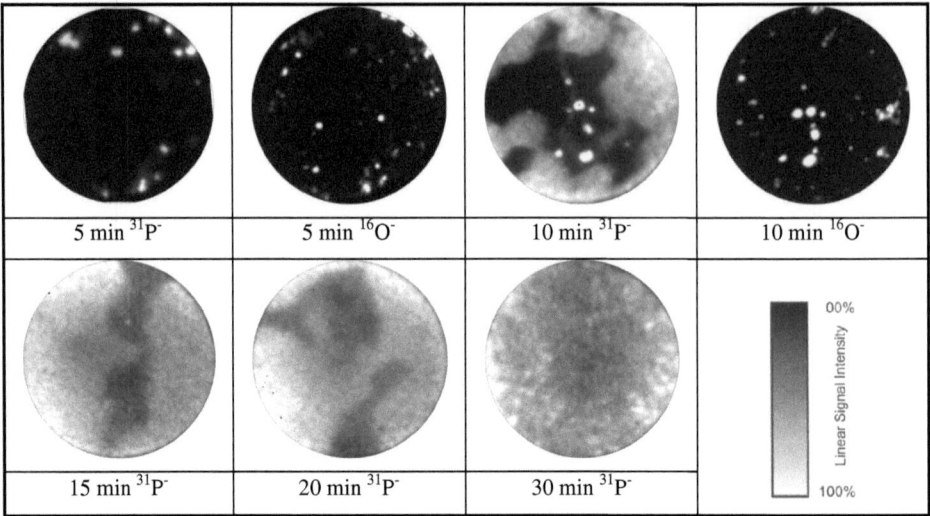

Figure 39: 2D SIMS images of $^{31}P^-$ and $^{16}O^-$ after well defined time in the sintering zone. The primary ions were Cs$^+$ at 14.5 keV. The primary ion beam was scanned over an area of 350 x 350 µm² with a primary ion current of 200 nA. The diameter of the images: 150 µm.

Figure 40 shows the almost homogonously P distribution independent on the used P amount. Phosphorus addition through Fe$_3$P powder causes the formation of a transient liquid phase by a eutectic reaction between Fe$_3$P and Fe during the sintering process at temperatures > 1050°C in protective atmosphere. The melt is then distributed by capillary forces. This liquid phase provides a better distribution of all other alloying components, a rounding of the pores and enhances diffusion in α – iron. The fractographic analysis made by means of Scanning Electron Microscopy on these samples showed the ductile fracture on the sample without any significant P content. Increasing P content leads the ductile behaviour of the fractures is decreased and finally the samples with 1wt% P show only intergranular fracture (insert in Figure 41). This intergranular fracture is an evidence for the grain boundary phosphorus segregation, but which couldn't be detected by means of SIMS. SIMS 2D analysis of the samples containing a increasing P amount show only a nearly homogenous P distribution with some brighter points corresponding to $^{30}Si^1H$. The real existence of P grain boundary segregation was detected by means of Auger electron spectroscopy (AES). Hereby the in – situ (inside the AES instrument) broken samples was investigated and the phosphorus coverage in the range of couple monolayers of the grain boundaries was found (Figure 41).

Results and discussion

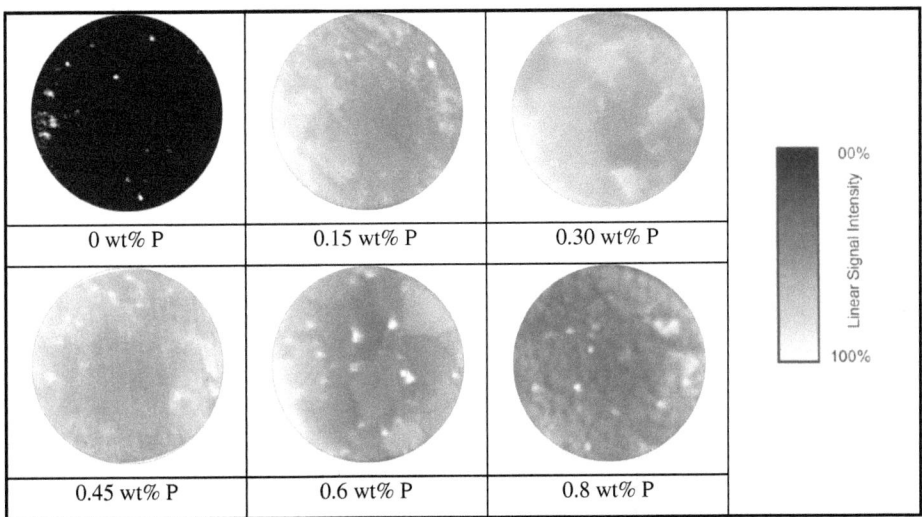

Figure 40: 2D SIMS images of $^{31}P^-$ after 60 min in the sintering zone with different amount on P. The primary ions were Cs$^+$ at 14.5 keV. The primary ion beam was scanned over an area of 350 x 350 µm² with a primary ion current of 200 nA. The diameter of the images: 150 µm.

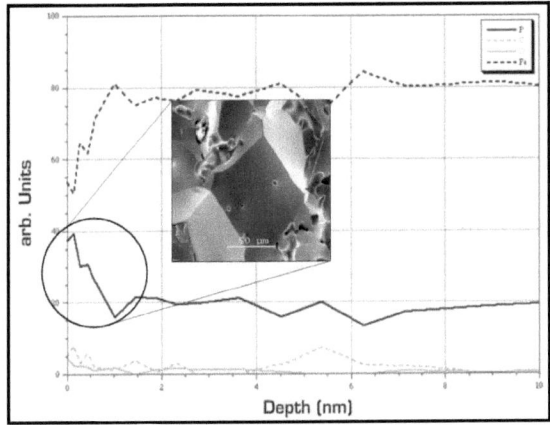

Figure 41: AES depth profile of in situ (vacuum) broken sample Fe+0.8wt%P. The insert shows the SEM image of the fracture on which the AES analysis was performed.

The second investigated sintering activator was boron. Here the influence of the used sintering atmosphere and the boron content will be demonstrated.

Results and discussion

If sintering is performed in protective atmosphere of flowing hydrogen, boron precipitates at grain boundaries are observable. At higher concentrations (0,15 wt% and above) besides boron precipitations a grain boundary network of boron and iron is formed. During the sintering process a liquid phase of Fe and Fe_2B is formed by eutectic reactions. Because of the very low solubility of boron in iron a grain boundary network of iron and boron remains (Figure 42 top). If nitrogen protective atmosphere is used, the formation of different nitrides and oxynitrides during sintering process is possible and no eutectic network was formed. The boron distribution is the same throughout the whole series, whereby only boron precipitates are observable. These completely different results were obtained due to the high affinity of boron to nitrogen (Figure 42 bottom).

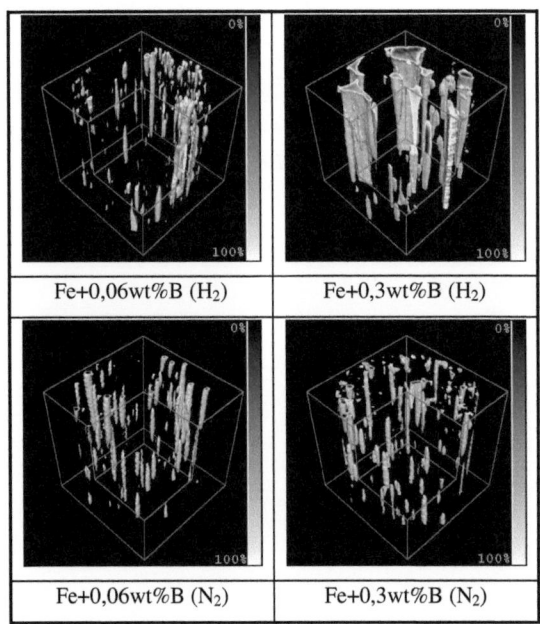

Figure 42: 3D SIMS images of boron distribution in isosurface mode of two different samples sintered in protective atmosphere of hydrogen (top) and nitrogen (botoom). Dimensions: lateral 150µm x 150µm, depth 5µm.

The further investigation results are presented in Appendix:
- 9.1 Phosphorus as Sintering Activator in Powder Metallurgical Steels: Characterization of the Distribution and its Technological Impact

Results and discussion

- 9.2 Characterization of the Distribution of the Sintering Activator Boron in Powder Metallurgical Steels with SIMS
- 9.3 SIMS Investigation of Cr – Mo Low Alloyed Steels Sintered With Boron
- 9.4 2D and 3D SIMS Investigations on sintered steels

5.2 Tribology

The ball – on – disk (BOD) test gives information on the material behaviour under pure sliding conditions and the rolling contact fatigue (RCF) tests were performed to characterise the reaction layer formation under rolling conditions.

The investigations after the BOD- and RDF- tests begun by means of optical profilometer in order to characterise the surface topography and to compare wear and friction behaviour. The material surface was also inspected by means of SEM and energy dispersive x – ray spectrometer (Figure 43).

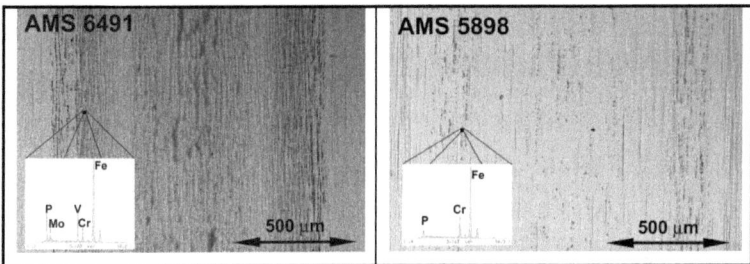

Figure 43: SEM images of the track on the rod surface (left...AMS 6491, right...AMS5898) after RCF tests. The rolling direction was perpendicular. The insets show the results of the EDX investigations made on the marked points.

Figure 43 shows the difference in the reaction layer formation of these two investigated materials. AMS 6491 exhibit a well developed reaction layer whereas almost no reaction layer on AMS 5898 can be observed. The reaction layer was then characterised more in detail by means of SIMS. The measurements were applied in the centre of the track as well as on the edge of the tracks. SIMS in depth profiles of P show a clearly difference for these two materials. AMS 6491 shows a high intensity of P for the first 38 min (approximately 200 nm), then a P enriched region up to 70 min sputtering (approximately 350 nm). AMS 5898 shows only for the first 4 min a well developed reaction layer followed by the phosphorus enriched region, again up to 70 min (depth ~350nm).

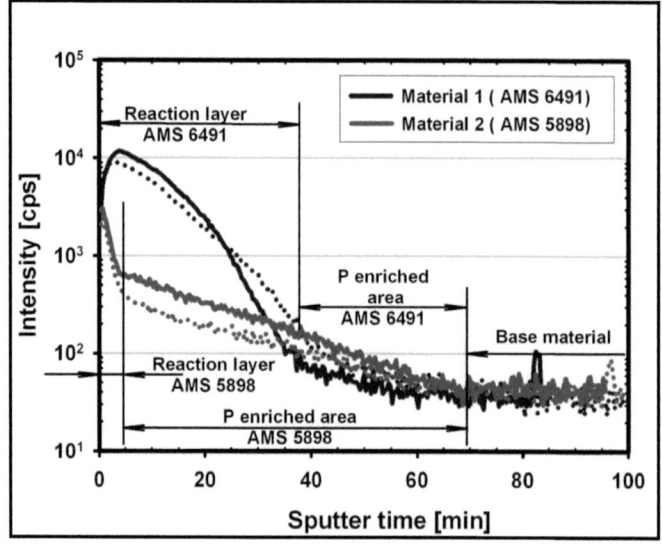

Figure 44: SIMS depth profiles of the investigated materials. The $^{31}P^-$ in depth distribution is demonstrated to be in three different regions. The primary ions were Cs^+ at 14.5 keV. The primary ion beam was scanned over an area of 350 x 350 µm² with a primary ion current of 10 nA. The secondary ions were then collected from the centre (Ø 30µm) of scanned area.

The 2D SIMS images of $^{31}P^-$ and of $^{31}PO_X^-$ made on both materials track surfaces shows the different distribution of 31P⁻ respectively of $^{31}PO_X^-$ clusters. Due to the same distribution of $^{31}P^-$ and $^{31}PO_X^-$ within one investigated sample it can be assumed that the formation layer is consisting of phosphates. The main difference between AMS 6491 and AMS 5898 is the distribution of the measured $^{31}P^-$ and $^{31}PO_X^-$. Thus it can be concluded that also the distribution of reaction layer is different. On AMS 6491 the reaction layer is well pronounced, whereas on AMS 5898 not. These results agree very well with those made by means of SEM and EDX as already shown in the Figure 43.

Results and discussion

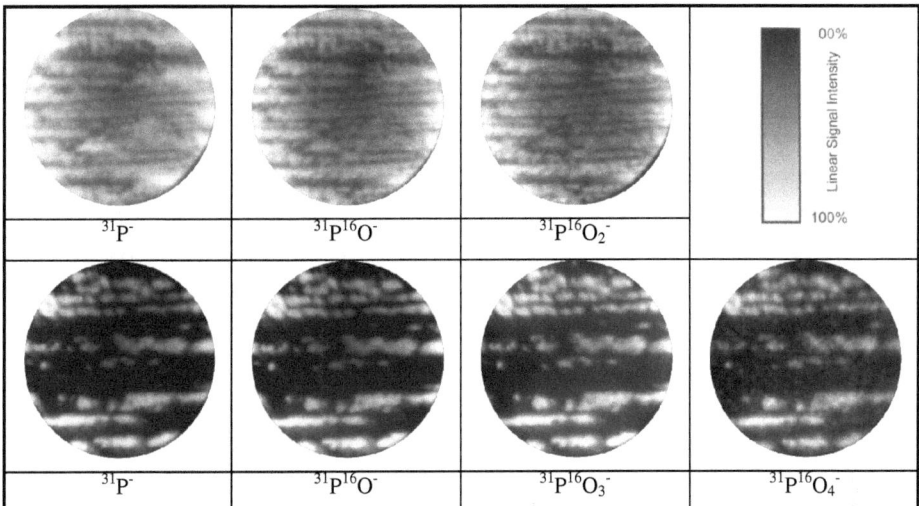

Figure 45: SIMS 2D images showing P⁻ and PO_X^- images on the surface of the investigated samples. The primary ions were Cs⁺ at 14.5 keV. The primary ion beam was scanned over an area of 350 x 350 µm² with a primary ion current of 10 nA. The diameter of the images: 150 µm.

The further investigation results are presented in Appendix:
- 9.5 Characterization of wear and surface reaction layer formation on aerospace bearing steel M50 and a nitrogen-alloyed stainless steel

5.3 Ion imlantation and defect engineering

5.3.1 Impurity gettering – gettering layer

From the measured depth profiles of the respective standards (copper implantation standard 1.4 MeV / $6*10^{13}$ at/cm² and phosphorus homogenous standard $4.5*10^{18}$ at/cm³), relative sensitivity factors (RSF) for copper (^{63}Cu) and phosphorus (^{31}P) regarding silicon (^{30}Si) as reference mass were calculated. With these sensitivity factors the quantification of impurity concentrations in the samples was possible. In addition to ^{63}Cu and ^{30}Si the elements ^{12}C and ^{16}O, which can also be gettered at the defect centres as well as the molecule ion ^{63}Cu^{28}Si, to get a second signal for the copper trend, were measured.

Results and discussion

The TRIM calculations for the 3.5MeV P⁺ ion implantation into silicon (note: TRIM calculations are made with amorphous silicon) predict a mean projected ion range for 2.7 µm. The P depth profile fits very well with the TRIM calculation. The SIMS measurements show two gettering layers formed after the implantation and subsequently annealing for 5 min at 900°C (Figure 46). The gettering in the Rp/2 – layer is much more pronounced for this annealing time as in the R_P – layer. No copper gettering is observed in the deeper regions than the mean projected range of the implanted ions. The defect free zone is similar to the silicon implanted sample #2 between 1.8 µm and 2.3 µm. In the Rp/2 gettering layer the profile maxima of Cu and O do not coincide. The C distribution is not affected (Figure 46).

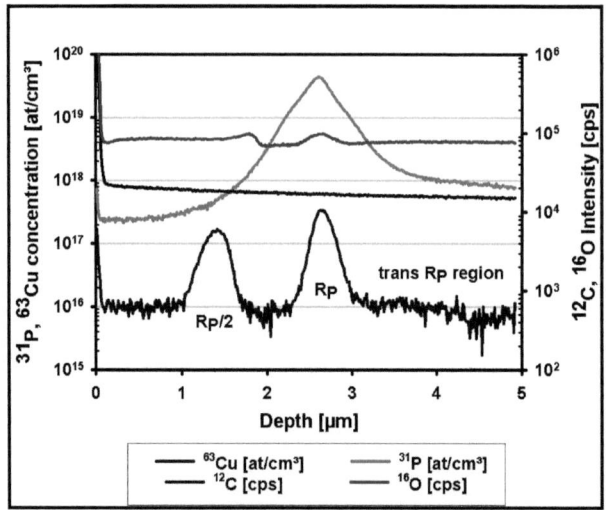

Figure 46: SIMS depth profiles of ^{63}Cu⁻, ^{12}C⁻, ^{31}P⁻ and ^{16}O⁻ in 3.5 MeV P⁺ (ion dose: 5*10^{15} atoms/cm²) implanted sample and subsequently annealed for 5 min at 900°C. The primary ions were Cs⁺ at 14.5 keV. The primary ion beam was scanned over an area of 300 x 300 µm² with a primary ion current of 150 nA. The secondary ions were then collected from the centre (Ø 60µm) of scanned area.

For the same experiment, but only with increased annealing time, 20 instead of 5 min, the gettering is pronounced now in three different layers: the Rp/2 – layer at 1.5 µm, the now dominating gettering in the Rp – layer at 2.8 µm and moreover impurity gettering in the trans – Rp – layer at approximately 4.3 µm (Figure 47). The formation respectively the tendency of impurity gettering in the regions beyond the mean projected ion range is only observed for P⁺ and As⁺ but not for Si⁺

implanted samples. The copper and oxygen gettering in the Rp – layer as demonstrated at approximately 2.8 µm in Figure 47 and at 2.7 µm in Figure 46 agrees very well with the measured P profile maximum. The P distribution is slightly broadened because of P diffusion during annealing.

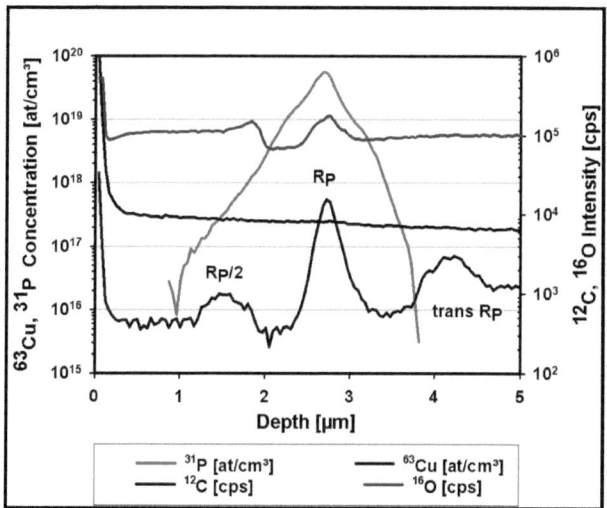

Figure 47: SIMS depth profiles of $^{63}Cu^-$, $^{12}C^-$, $^{31}P^-$ and $^{16}O^-$ in 3.5 MeV P$^+$ (ion dose: $5*10^{15}$ atoms/cm²) implanted sample and subsequently annealed for 20 min at 900°C. The primary ions were Cs$^+$ at 14.5 keV. The primary ion beam was scanned over an area of 300 x 300 µm² with a primary ion current of 150 nA. The secondary ions were then collected from the centre (Ø 60µm) of scanned area.

Figure 48 demonstrates different gettering behaviour if P$^+$ ions are implanted with the same implantation energy of 3.5 MeV but with lower dose $5*10^{14}$ at/cm² (instead of $5*10^{15}$ at/cm²) and annealed for 5 min. As expected the P concentration is one magnitude lower at the maximum, due to the lower ion fluency, but unexpected only small impurity gettering is reached in the Rp/2 layer at 1.5 µm, relatively weak Cu gettering appears at the projected ion range at 2.6 µm, but also unexpected the dominating copper gettering is observed at the region beyond (deeper than 3.6 µm) the projected ion range in the trans – Rp gettering range. Oxygen is mobilized by the annealing step and its trapping is diffusion limited. The oxygen in depth distribution shows here no gettering neither in the Rp/2 layer nor in the region of the Rp – layer.

Results and discussion

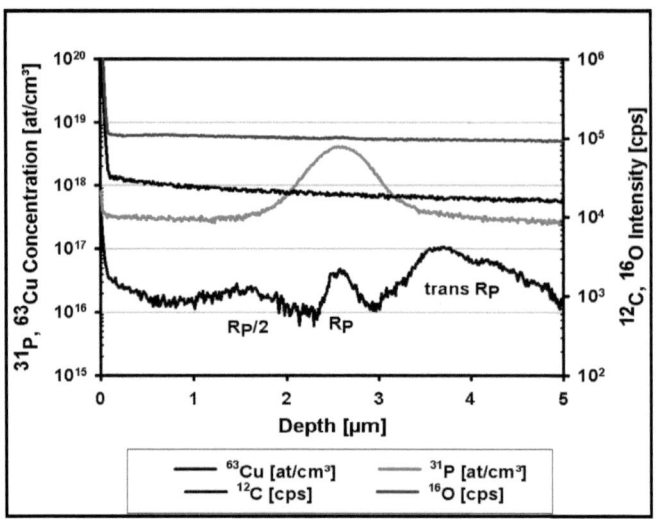

Figure 48: SIMS depth profiles of $^{63}Cu^-$, $^{12}C^-$, $^{31}P^-$ and $^{16}O^-$ in 3.5 MeV P$^+$ (ion dose: 5*10^{14} atoms/cm²) implanted sample and subsequently annealed for 5 min at 900°C. The primary ions were Cs$^+$ at 14.5 keV. The primary ion beam was scanned over an area of 300 x 300 µm² with a primary ion current of 150 nA. The secondary ions were then collected from the centre (Ø 60µm) of scanned area.

5.3.2 Defect engineering for ion beam synthesis of SOI structures

In this part different defect engineering methods are applied to study the formation of SiO$_2$ formation in Si by means of ion beam synthesis. The typical oxygen ion implantation (200 keV, $1*10^{17}$ at/cm²) into silicon as implanted and annealed is shown in the Figure 49. The in depth distribution of oxygen of the annealed sample is due to the thermal activated diffusion of oxygen slightly broaden.

Results and discussion

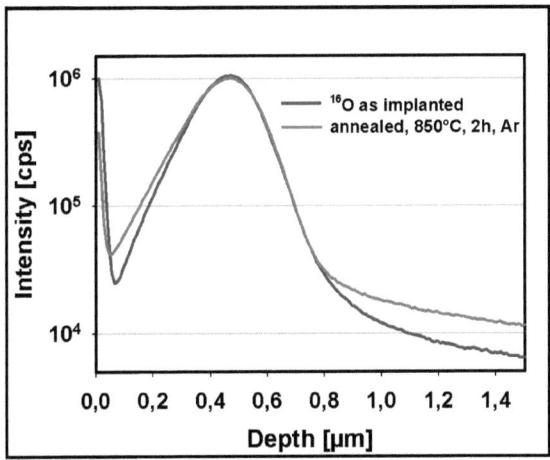

Figure 49: Oxygen SIMS depth profile of 200 keV O (ion dose: $1*10^{17}$ atoms/cm²) as implanted and annealed sample. The primary ion beam (Cs@14.5keV) was scanned over an area of 300 x 300 µm² with a primary ion current of 150 nA. The secondary ions were then collected from the centre (Ø 60µm) of scanned area.

At first step of the defect engineering the pre – deposition of He cavities by means of He pre – implantation was made. These should act as trapping centres for O and to allow the volume expansion for the formation of SiO_2. Figure 50 (top right) shows a TEM image of generated He bubbles and (bottom right) of SiO_2 precipitates (SiO_2 filled cavities, TEM in underfocussed mode) after subsequently simultaneous dual O and Si implantation. This method is promising to be better for ion beam synthesis for implantation temperatures above 400°C. The pre deposition of He cavities as well as the subsequently implantation of Si (in situ generation of excess vacancies) and O lead to the formation of narrower SiO_2 layer and improves the crystal quality of Si around SiO_2. The further result of the above described defect engineering is described with SIMS depth profiles of the sample with as implanted O and of the sample with O after the above described prior and in situ defect engineering. The small but significant enrichment of O at the maximum of the implantation range is clearly visible (Figure 50). This enrichment occurs directly at the mean projected range of O and He which is here optimized to be at 440 nm.

Results and discussion

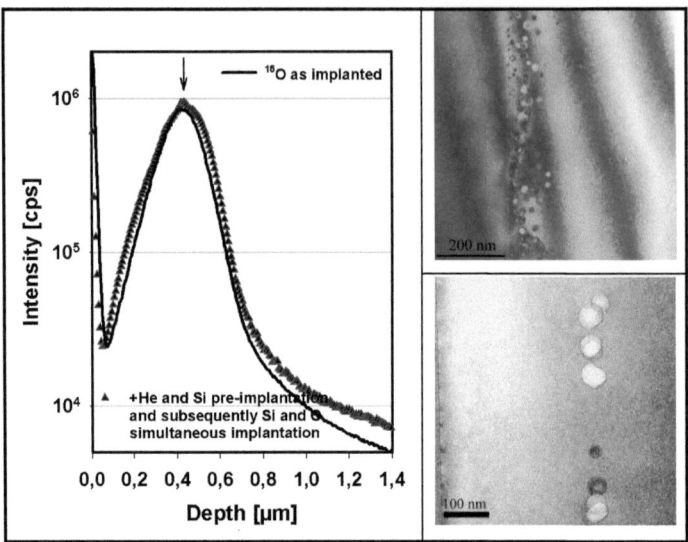

Figure 50: Left: Oxygen SIMS depth profile of as implanted sample (200 keV O, ion dose: $1*10^{17}$ atoms/cm²) and of sample with He (45 keV, ion dose: $4*10^{16}$ atoms/cm²) and Si (1 MeV, ion dose: $3.5*10^{15}$ atoms/cm²) pre – implantation and subsequently Si und O simultaneous implantation. The primary ion beam (Cs@14.5keV) was scanned over an area of 300 x 300 µm² with a primary ion current of 150 nA. The secondary ions were then collected from the centre (Ø 60µm) of scanned area. Right: (top) TEM image of He bubbled, introduced by He implantation and (bottom) TEM image of SiO_2 precipitates after He pre – implantation and subsequently O and Si simultaneous implantation.

In addition to pre deposition of gettering centres the used oxygen ion dose which is necessary for production of buried SiO_2 layer can be partly reduced by well defined surface oxidation. Figure 51 shows the in diffusion of one part of the surface O after the controlled (7min, 1150°C, wet atmosphere) oxidation step and subsequently removal of the oxidation layer.

Results and discussion

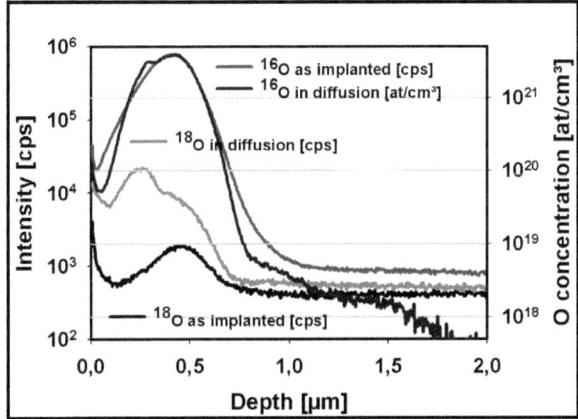

Figure 51: SIMS depth profiles of $^{16}O^-$ and $^{18}O^-$ in 200 keV O (ion dose: $1*10^{17}$ atoms/cm²) as implanted and oxidised (7 min, wet atmosphere, 1150°C) sample. The primary ion beam (Cs@14.5keV) was scanned over an area of 300 x 300 µm² with a primary ion current of 150 nA. The secondary ions were then collected from the centre (Ø 60µm) of scanned area.

The combination of the three above mentioned engineering steps: introduction of He vacancies, simultaneous dual Si and O implantation and oxygen in diffusion over well controlled surface oxidation; which is very promising for the ion beam synthesis of SOI structures is shown in

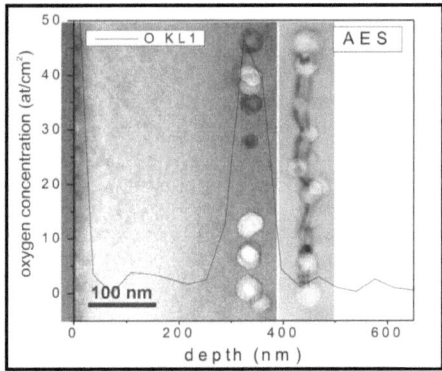

Figure 52: AES oxygen depth profile of sample with prior introduction of He cavities and subsequently dual implantation of Si and O. Sample was annealed for 5h at 1250°C (BOX anneal). TEM image (left) shows SiO_2 filled cavities and almost defect free Si around SiO_2. The insert (TEM image on right) shows strain contrast between SiO_2 precipitates.

Results and discussion

The further investigation results are presented in Appendix:
- 9.6 SIMS Investigation of Gettering Centres Produced by Phosphorus MeV Ion Implantation
- 9.7 Investigation of Gettering Effects in CZ-type Silicon with SIMS
- 9.8 Study of defect engineering in the initial stage of SIMOX processing

5.4 SiGe Semiconductors and Heterostructures

The complete results of this investigation part are presented in chapter 9.9 (Appendix) and chapter 9.10 (Appendix). The highlights are shown here.

The $Si_{1-x}Ge_x$ samples as described in chapter 4.4 were at first investigated by means of low energy RBS. Additionally SIMNRA calculations, considering non – Rutherford cross sections, isotope effects, realistic stopping powers, energy loss straggling, surface roughness, dual and multiple scattering, were parallel performed to compare these with measured spectra. Figure 53 shows both simulated and measured spectra for sample containing 5 at% Ge in the quantum well. The Si cap layer is thicker than expected, and therefore the measured position of the Ge high energy edge is at lower energies. The composition of the individual layers is optimized such that the plateaus due to scattering from Si and from Ge both are reproduced by the SIMNRA simulation. The concentration of doping element antimony was lower than the detection limit of the method, thus there is no contribution neither in the simulated, nor in the measured spectrum.

Results and discussion

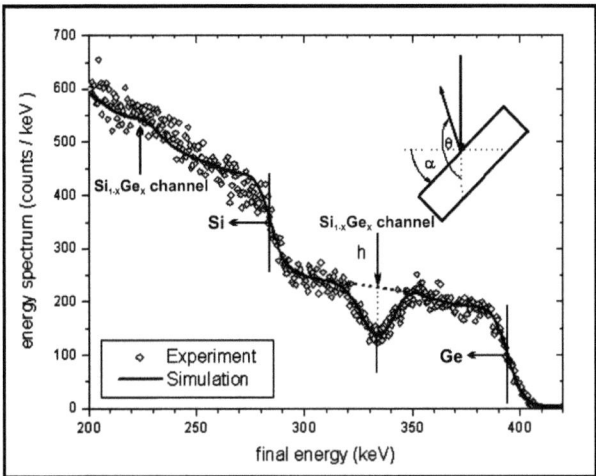

Figure 53: Measured (open diamonds) and SIMNRA (line) simulated low energy RBS depth profile of sample with $Si_{1-X}Ge_X$ (x: 0.05) composition in the quantum well.

In contrary Figure 54 shows the depth profiles for $^{30}Si^+$ and $^{74}Ge^+$ of all measured samples. It can be seen, that even the sample containing 0 at% Ge does not show a sufficient signal decrease in the quantum well. The quantum well interfaces should appear as sharp edges, but due to the problem of the method the profile is distorted by the primary beam. Anyway the measurement of such samples is a challenge for SIMS because the layer thickness, here the quantum well, is within the depth resolution of the method. This depth profile distortion can be explained as a convolution of the real nature of the sample with a SIMS response function.

Results and discussion

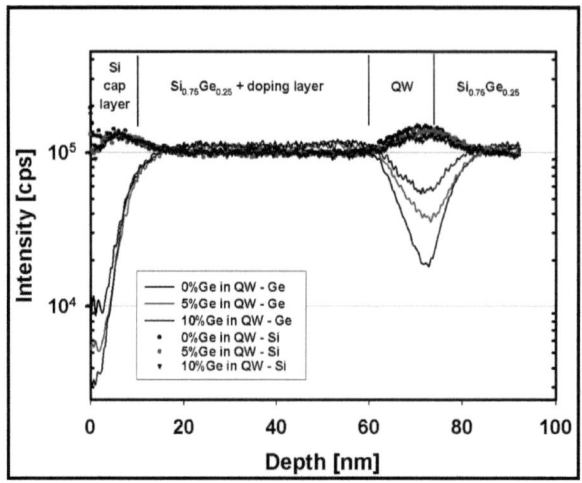

Figure 54: SIMS ^{30}Si$^+$ and ^{74}Ge$^+$ depth profiles of the investigated samples. The used primary ions were O$_2^+$ with energy of 5.5 keV. The primary ion beam was scanned over an area of 350 x 350 µm² with a primary ion current of 30 nA, secondary ions were collected from the centre (Ø 60µm) of scanned area.

Measured and normalised depth profile of one well defined structure e.g. Ge δ layer (*one* monolayer of Ge deposited by MBE with a cap layer of Si also deposited by MBE) is used as the response function of the SIMS instrument. Figure 56 shows SIMS depth profile and schematic view of Ge δ layer, which is used for further calculation of the Ge concentration, position and thickness of the quantum well. Now inverse modelling of as already shown distorted profiles was made by convolving an assumed undistorted profile (rectangle model) point to point with the response function. This procedure is demonstrated in the Figure 55 and Equation 8. The resulting convolution profile was manually fitted to the measured Ge profile varying the following parameters of the used model: Ge intensity before, after and in the quantum well, position and width of the quantum well; until the best match of both graphs was reached. The obtained results are shown in the Figures 53 – 55.

```
For x = 1 to a
Y = 0
    For y = 1 to b
    Y = Value [model, (a)+(b)] * Value [response function, (b+1)-(b)]
    Next b
Next a                                                        Equation 8
```

Results and discussion

x   Outer (first) loop with running variable a
a   Number of points to be calculated for the convolution
y   Inner (second) loop with running variable b
b   Number of valued the response function is consisting of
Y   Result for each convoluted point

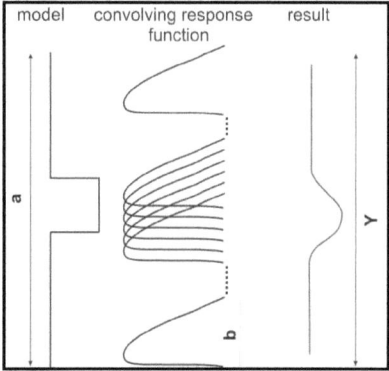

Figure 55: Schematic overview of the point to point forward convolution.

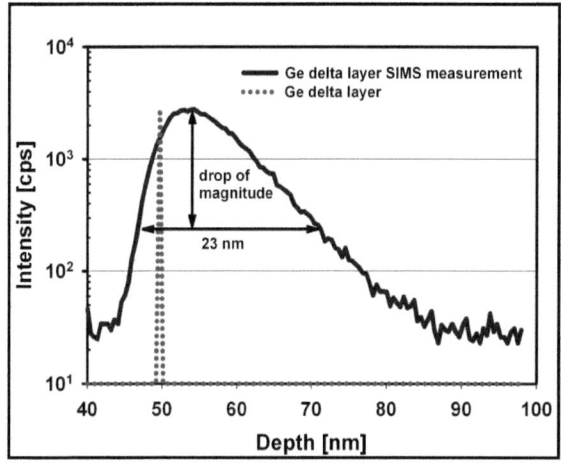

Figure 56: SIMS depth profile and the schematic view of the Ge δ layer.

87

Results and discussion

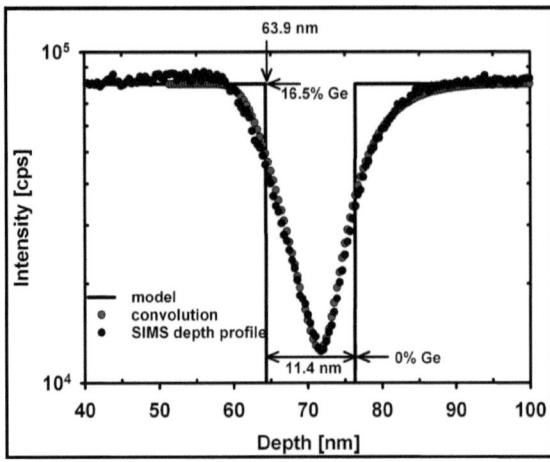

Figure 57: Convolution procedure for sample with nominally 0%Ge in the quantum well. The used rectangle model, the measured SIMS depth profile and the convolution product are demonstrated.

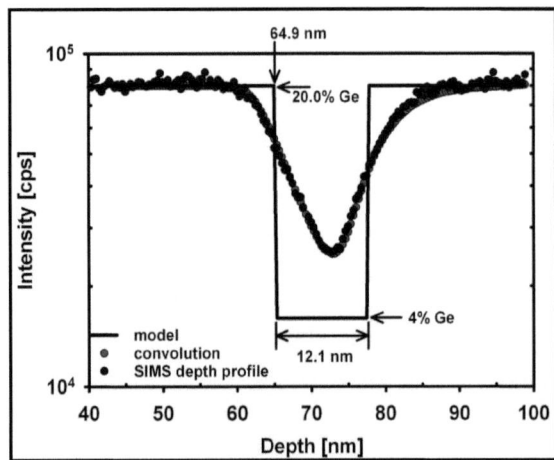

Figure 58: Convolution procedure for sample with nominally 5%Ge in the quantum well. The used rectangle model, the measured SIMS depth profile and the convolution product are demonstrated.

Results and discussion

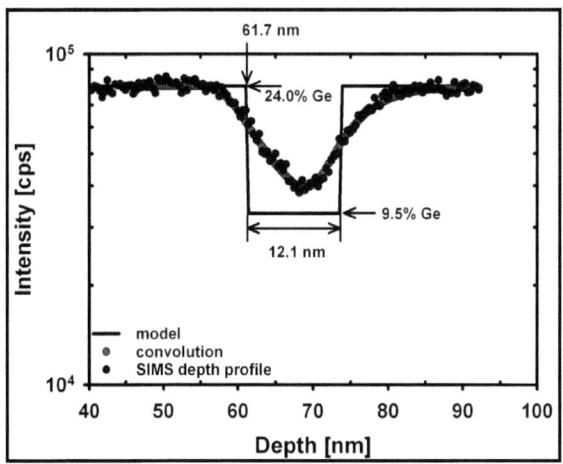

Figure 59: Convolution procedure for sample with nominally 10%Ge in the quantum well. The used rectangle model, the measured SIMS depth profile and the convolution product are demonstrated.

In the Figure 53 one RBS depth profile of the investigated samples was shown, but as above mentioned no contribution of doping element antimony to the profile could be demonstrated. Using SIMS, there was no problem to identify the in depth distribution of Sb (Figure 60).

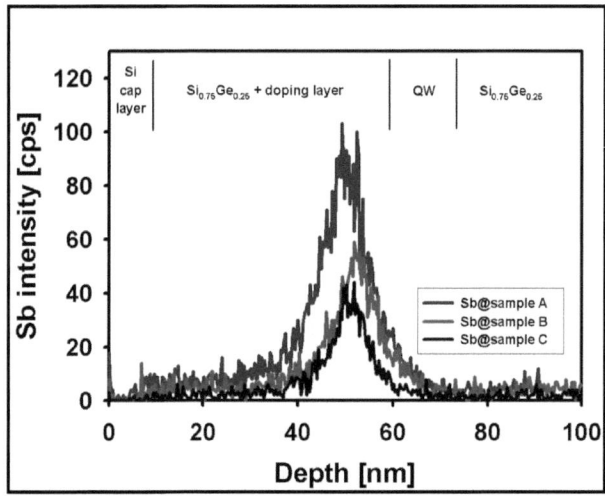

Figure 60: SIMS depth profile of the doping element Sb. The used primary ions were O_2^+ with energy of 5.5 keV. The primary ion beam was scanned over an area of 350 x 350 μm² with a primary ion current of 100 nA, secondary ions were collected from the centre (Ø 60μm) of scanned area.

The further investigation results are presented in Appendix:

Results and discussion

- 9.9 Low energy RBS and SIMS analysis of the SiGe quantum well
- 9.10 Quantitative Analysis of the Ge Concentration in a SiGe Quantum Well: Comparison of Low Energy RBS and SIMS Measurements

6 Conclusion

The use of the physical analytic methods in materials science is essential. It could be shown that due to their figures of merit, SIMS, AES, RBS and EM are complementary methods each one with advantages but also with some drawbacks.

For SIMS it was not possible to determine the monolayer phosphorus grain boundary segregations, but it was the only method able to detect 2D and 3D element or cluster distribution for sintering activators as well as for impurities and thus to contribute a big part for the explanation of investigated sintering processes and the influence of the used sintering activator and alloying elements. SEM was the method of choice for investigating the fracture behaviour of studied samples.

Beside SEM, SIMS was very helpful to characterise the reaction layer formation more in detail. By means of depth profiling and 2D SIMS the different reaction layer formation on the two investigated materials was demonstrated. Due to the comparison of all obtained SIMS results (2D element distribution and dept profiles) it is assumed that the reaction layer, formed during the tribological test, is consisting of phosphates. This is a problem using dynamic SIMS as demonstrated here. This problem can be solved by means of static SIMS, e.g. time of flight SIMS, whereby the monitoring of such clusters and not only elements is improved.

SIMS is not only a power tool using it for 2D and 3D elemental distribution but also for depth profiles of trace elements with the detection limit in the range of at least 1 ppm. It was shown on the example of the ion implantation into silicon for the generation of the gettering layer that the quantification of copper and phosphorus with suitable implantation standards strengthens the superiority of SIMS versus other methods for this kind of problems. The investigation of the defects produced by ion implantation and subsequently annealing was supported by TEM. The defects at around mean projected range, R_P, of the implanted ions, and in the region between surface and R_P could be detected by means of cross section TEM, whereby a special specimen preparation was needed. The detection of the gettering layer formed beyond the R_P, trans R_P, was only successful by means of SIMS.

SIMS, AES and TEM were also applied for the detection of the oxygen concentration and respectively for the detection of He induced cavities at the defect engineering for ion beam

Conclusion

synthesis of silicon on insulator (SOI) structures. The oxygen depth profiles and its concentration, which is changed (increasing) from as implanted state to the state after annealing and He pre implantation (cavities), and as well as the results after the controlled surface oxidation was helpful to describe that the defect engineering is promising technique for ion beam synthesis of SOI structures but it needs some improvements and optimisation.

The correspondence, the ability and the limits of SIMS and RBS were demonstrated on the example of Ge quantitative analysis inside SiGe quantum wells. RBS has shown its outstanding opportunity of standard free quantification and sufficiently good depth resolution and depth profiling in the low energy mode. Nevertheless the detection of doping element antimony here was not possible at first because of the used mode but also because of the detection limit of RBS. SIMS was able to detect the in depth distribution of doping element antimony and thus demonstrated its indispensable role at semiconductor industry. Sometimes the depth profile of desired elements is sufficient and it is not always necessary to calculate the concentration of this element out of the depth profile but nevertheless, here also one well defined implantation standard would be needed to calculate the antimony concentration out of its measured depth profile. The problem of the limited depth resolution of SIMS was also here demonstrated and it was shown that using mathematical models, e.g. point to point forward convolution, it is possible to improve the depth resolution of the used SIMS instrument. The microelectronic devices produced nowadays are permanently getting smaller and the use of SIMS with lower primary ion energy is possible but here some problems of beam stability, long time measurements and beam geometry are present thus the work with mathematical models is necessary

There is no denying that the use of more physical analytical methods in the materials science is essential. In materials science and research the correspondence of these techniques as well as the correspondence and the cooperation of the people using and studying its results is not less important (Figure 61).

Conclusion

Figure 61: Cooperation pays[75].

Conclusion

7 References

[1] German, Randall M.: Powder metallurgy of iron and steel; New York, Weinheim: JOHN WILEY & SONS, INC., 1998

[2] Schatt, W., Wieters, K. P., Pulvermetallurgie; Düsseldorf: VDI Verlag, 1994

[3] Gmelin – Durrer: Mettalurgy of Iron 4, Vol. 10.; Berlin, Heidelberg, New York: Springer, 1992,

[4] Thümmler, F., Oberacker, R.: Introduction to powder metallurgy; Book 490; London: The Institute of Materials, 1993

[5] L'Estrade, L., Hallen, H.: Production of water- and gas- atomized powders Hoeganaes AB; Hoeganaes, Sweden; ATB Metallurgie (1987)

[6] Höganäs Eisen und Stahlpulver für Sinterformteile; Höganäs 1998

[7] Danninger H.: Pulvermetallurgie Beispiele für den Produkteinsatz; Wien: WIFI Österreich, 1996

[8] Molinari, A.; Straffelini, G.; Fontanari, V.; Canteri, R.: Sintering and microstructure of phosphorus steels; Powder Metallurgy (1992), 35(4), 285-91

[9] Molinari, A.; Straffelini, G.; Canteri, R.: Heat treatment and mechanical behaviour of sintered Fe-C-P steels; International Journal of Powder Metallurgy (Princeton, New Jersey) (1994), 30(3)

[10] Molinari, A.; Kazior, J.; Marchetti, F.; Canteri, R.; Cristofolini, I.; Tiziani, A.: Sintering mechanisms of boron alloyed AISI 316L stainless steel; Powder Metallurgy (1994), 37(2), 115-22

[11] Online: GKN Sinter Metals: http://www.gknsintermetals.com/

[12] DIN 50 323, Teil 1: Tribologie; Begriffe. Beuth Verlag, Berlin, (1988).

[13] Berns, H., Ebert, F.J., and Zoch, H.-W., "The New Low Nitrogen Steel LNS – A Material for Advanced Aircraft Engine and Aerospace Bearing Applications", Bearing steels: Into the 21st century, ASTM STP 1327, J.J.C.Hoo and W.B. Green; Eds., American Society for Testing Materials, West Conshohocken, PA, 1998, pp. 354-373

[14] Godfrey,D, The lubrication mechanism of tricresyl phosphate on steel, ASLE Trans., 8 (1) (1965) 1

[15] Cornell, R.M., Schwertmann, U., "The Iron Oxides; Structure, Properties, Reactions, Occurrence and Uses", Weinheim – VHC, pp. 445-461, (1996)

[16] Winder,C., Balouet, J.-C, "The Toxicity of Commercial Jet Oils", Environmental Research Section A 89, pp. 146-164, (2002)

[17] Stachowiak, G.W. Batchelor, A.W. "Engineering Tribology", ISBN 0-7506-7304-4, pp.84. (2001)

[18] Saba, C.S. and Forster N.H., "Reactions of aromatic phosphate esters with metals and their oxides" Tribology Letters Vol. 12, No. 2, pp. 135, (2002)

[19] Berns, H., and Trojahn, W., "High-Nitrogen Cr-Mo Steels for Corrosion Resistant Bearings", Creative Use of bearing steels, ASTM STP 1195, J.J.C. Hoo, Ed., American Society for Testing Materials, West Conshohocken, PA, 1993, pp. 149-155

[20] Boehmer, H.J., Hirsch, T., and Streit, E., "Rolling Contact Fatigue Behavior of Heat Resistant Bearing steels at High Operational Temperatures", Bearing steels: Into the 21st century, ASTM STP 1327, J.J.C.Hoo and W.B. Green; Eds., American Society for Testing Materials, West Conshohocken, PA, 1998, pp. 131-151

[21] Trojahn, W., Streit, E., Chin, H., and Ehlert, D., "Progress in Bearing Performance and Advanced Nitrogen Alloyed Stainless Steel, Cronidur 30", Bearing steels: Into the 21st century, ASTM STP 1327, J.J.C.Hoo and W.B. Green; Eds., American Society for Testing Materials, West Conshohocken, PA, 1998, pp. 447-459

[22] Glover, D, "A ball-.rod contact fatigue tester", Rolling Contact Fatigue Testing of Bearing Steel, ASTM STP 771, J.J.C. Hoo, Ed., American Society for testing and Materials, 1982, pp. 107-127

References

[23] Y Wang and M Hadfield: Failure modes of ceramic rolling elements with surface crack defects, Wear, Vol. 256, pp.208-219, 2004

[24] Online: http://www.csm-instruments.com/

[25] F. Thuselt: Physik der Halbleiterbauelemente, Springer, Berlin (2004)

[26] Peter Y. Yu, Manuel Cardona: Fundamentals of Semiconductors, Springer, Berlib (2005)

[27] Online: http://www.ioffe.rssi.ru/SVA/NSM/Semicond/SiGe/bandstr.html

[28] US patent 2787564, Bell Laboratories, (1954)

[29] Walter Scott Ruska, Microelectronic processing, McGraw-Hill Companies (1986)

[30] The Stopping and Range of Ions in Solids", by J. F. Ziegler, J. P. Biersack and U. Littmark, Pergamon Press, New York, 1985 (new edition in 1999)

[31] Online: SRIM http://www.srim.org

[32] J. P. Biersack, Radiat. Eff. Defects Solids, 146, (1998), 27-48

[33] J. P. Biersack, Nucl. Instrum. Methods Phys. Res. B, 153, (1999), 398-409

[34] J. F. Ziegler, J. P. Biersack, U. Littmark, The Stopping and Range of Ions in Sol¬ids, Vol. 1, Pergamon Press, New York 1985

[35] Seidel, T. E. Gettering in Silicon. In Material Issues in Silicon Integrated Circuit Processing, Materials Research Society, Pittsburgh, Wittmer, M., Stimmell, J., Strathman, M., Eds.; 1986; 3 – 12

[36] H.Wong, N.W. Cheung, P.K. Chu, J.Liu and J.W. Mayer; Appl. Phys. Lett. 52, 1023 (1998)

[37] M. Tamura, T. Ando, K. Ohya; Nucl. Instr. Meth. Res. B, 59 – 60 (1991), 572

[38] R.A. Brown, O. Kononchuk, G.A. Rozgonyi, S. Koveshnikov, A.P. Knights, P.J. Simpson and F. Gonzalez: J. Appl. Phys. 84, 2459, (1998)

[39] V.C. Venezia, D.J. Eaglesham, T.E. Haynes, A. Agarwal, D.C. Jacobson, H.-J. Gossmann and F.H. Baumann: Appl. Phys. Lett. 73, 2980, (1998)

[40] O.W. Holland, L. Xie, B. Nielsen and D.S. Zhou: J. Electronic Mat. 25 - 1, 99, (1996)

[41] R. Koegler, A. Peeva, A. Lebedev, M. Posselt, W. Skorupa, G. Oezelt, H. Hutter, M. Behar: J. Appl. Phys. 94, 3834 (2003)

[42] A. Peeva, R. Koegler, W. Skorupa: Nucl. Instr. Meth. Res. B, 206 (2003) 71-75

[43] R. Koegler, A. Peeva, W. Anwand, G. Brauer, P. Werner, U. Gösele: Appl. Phys. Lett. 75/9 (1999), 1279 - 1281

[44] R. Koegler, A. Peeva, W. Anwand, P. Werner, A.B. Danilin, W. Skorupa: Solid State Phenomena 69-70 (1999) 235

[45] A. Peeva, R. Koegler, G. Brauer, P. Werner, W. Skorupa: Mmaterial Science in Semiconductor Processing, 3 (2000), 297 – 301

[46] Y. M. Gueorgiev, R. Koegler, A. Peeva, D. Panknin, A. Muecklich, R. A. Yankov, W. Skorupa: Appl. Phys. Lett. 75, 3467 (1999)

[47] Y. M. Gueorgiev, R. Koegler, A. Peeva, A. Muecklich, D. Panknin, R. A. Yankov, W. Skorupa: J. Appl. Phys. 88, 5465 (2000)

[48] J.R. Kaschny, R. Kögler, H. Tyrrof, W. Bürger, F. Eichhorn, A. Mücklich, C. Serre, W. Skorupa: Nuclear Instruments and Methods in Physics Research Section A 551/2-3, 200 – 207, (2005)

[49] Online: http://www.ioffe.rssi.ru/SVA/NSM/Semicond/SiGe/bandstr.html

[50] J. M. Fernández, L. Hart, X. M. Zhang, M. H. Xie, J. Zhang and B. A. Joyce: Journal of Crystal Growth 164 (1996) 241

References

[51] M. Maier, D. Serries, T. Geppert, K. Köhler, H. Güllich and N. Herres: Applied surface science 203-204 (2003), 486

[52] John H. Davies: The physics of low-dimensional semiconductors : an introduction, Cambridge Univ. Press, XVIII (2005)

[53] A. Cho, "Film Deposition by Molecular Beam Techniques," J. Vac. Sci. Tech., Vol. 8, S31-S38, (1971)

[54] A. Cho, J. Arthur, "Molecular Beam Epitaxy," Prog. Solid-State Chem., Vol. 10, 157-192, 1975

[55] Online: http://www.wsi.tu-muenchen.de/E24/resources/facilities.htm

[56] Online: http://people.deas.harvard.edu/~jones/ap216/images/bandgap_engineering/bandgap_engineering.html

[57] H. Hutter, "Dynamic Secondary Ion Mass Spectrometry"; H. Bubert and H. Jenett, Surface and Thin Film Analysis, Wiley – VCH, pp. 106-121, (2002)

[58] Charles Evans Online Tutorials: http://www.cea.com

[59] F. K. Herzog, R. Viehböck, Phys. Rev., 76, 855, (1949)

[60] Grasserbauer M., Dudek H. J., Ebel M. F.: Angewandte Oberflächenanalyse, Springer Verlag (1985)

[61] Lareau, T. Richard, M. Wood, F. Chmara, Inst. Phys.Conf. Ser., 165 (Microbeam Analysis 2000), (2000), 329 – 330

[62] Piplits K., Tomischko W., Brunner C. H: Proceedings of SIMS X, Wiley, (1995)

[63] C. R. Brundle: Encyclopedia of Materials Characterization, Butterworth-Heinemann 1992

[64] M. Mayer: Ion beam analysis of rough thin films, Nucl. Instrum. Methods Phys. Res. B194 (2002) 177

[65] S. Humphries: Principles of Charged Particle Acceleration, John Wiley and Sons, New York 1986

[66] Lawrence R. Doolittle: Nucl. Instrum. Meth. B, 9/3 (1985) 344

[67] Geoffrey A. Meek: Practical Electron Microscopy, Wiley 1978

[68] Online: http://www.matter.org/tem/default.htm

[69] Online: http://en.wikipedia.org/wiki/Main_Page

[70] Online: http://www.mete.metu.edu.tr/Facilities/Service/sem/Semlab.htm

[71] T. Heiser, A Mesli: Appl. Phys. A 57 (1993) 325-328

[72] Schäffler F., Többen D., Herzog H-J., Abstreiter G., Holländer B.: High-electron-mobility Si/SiGe heterostructures: influence of the relaxed SiGe buffer layer, Semicond. Sci. Technol. 7 (1992) 260

[73] F. Kastner: PhD thesis, Johannes Kepler Universität Linz

[74] Kasper E., Kibbel H., Schäffler F.: An Industrial Single Slice Si-MBE Apparatus, Journal of Electrochemical Society 136 (1989) 1154

[75] Helmut W. Werner: SIMS: from research to production control; Surface and Interface Analysis 35, (2003), 859

8 Table of figures

Figure 1: Online: GKN Sinter Metals: http://www.gknsintermetals.com/

Figure 2: ASTM Special Technical Publication (2007), STP 1465(Bearing Steel Technology: 7th Volume), 214-223

Figure 3: Online: http://www.csm-instruments.com/

Figure 5: Borchardt-Ott, Walter, Kristallographie, Springer Verlag, Berlin (1993)

Figure 6: Peter Y. Yu, Manuel Cardona: Fundamentals of Semiconductors, Springer, Berlib (2005)

Figure 7: Walter Scott Ruska, Microelectronic processing, McGraw-Hill Companies (1986)

Figure 9: Nuclear Instruments&Methods in Physics Research, Section B: Beam Interactions with Materials and Atoms (2005), 240, (3), 733-740

Figure 10: John H. Davies: The physics of low-dimensional semiconductors: an introduction, Cambridge Univ. Press, XVIII (2005)

Figure 11: Online: http://people.deas.harvard.edu/~jones/ap216/ images/ bandgap _engineering/bandgap_engineering.html

Figure 14, 18 - 28, 30, 34 - 37: Online http://www.cea.com

Figure 15: Author Prof. Herbert Hutter, Vienna University of Technology

Figure 17: Online http://www.cameca.fr

Figure 29: C. R. Brundle: Encyclopedia of Materials Characterization, Butterworth-Heinemann 1992

Figure 31 - 32: Geoffrey A. Meek: Practical Electron Microscopy, Wiley 1978

Figure 33: Online: http://www.mete.metu.edu.tr/Facilities/Service/sem/Semlab.htm

Figure 61: Helmut W. Werner: SIMS: from research to production control; Surface and Interface Analysis 35, (2003), 859

9 Appendix

9.1 Phosphorus as Sintering Activator in Powder Metallurgical Steels: Characterization of the Distribution and its Technological Impact

9.2 Characterization of the Distribution of the Sintering Activator Boron in Powder Metallurgical Steels with SIMS

9.3 SIMS Investigation of Cr – Mo Low Alloyed Steels Sintered With Boron

9.4 2D and 3D SIMS Investigations on sintered steels

9.5 Characterization of wear and surface reaction layer formation on aerospace bearing steel M50 and a nitrogen-alloyed stainless steel

9.6 SIMS Investigation of Gettering Centres Produced by Phosphorus MeV Ion Implantation

9.7 Investigation of Gettering Effects in CZ-type Silicon with SIMS

9.8 Study of defect engineering in the initial stage of SIMOX processing

9.9 Low energy RBS and SIMS analysis of the SiGe quantum well

9.10 Quantitative Analysis of the Ge Concentration in a SiGe Quantum Well: Comparison of Low Energy RBS and SIMS Measurements

9.1 Phosphorus as Sintering Activator in Powder Metallurgical Steels: Characterization of the Distribution and its Technological Impact

D. Krecar, V. Vassileva, H. Danninger, H. Hutter

Published in: Analytical and Bioanalytical Chemistry, Springer

Volume 379, Number 4, 2004

610 – 618

SPECIAL ISSUE PAPER

Dragan Krecar · Vassilka Vassileva · Herbert Danninger
Herbert Hutter

Phosphorus as sintering activator in powder metallurgical steels: characterization of the distribution and its technological impact

Received: 14 November 2003 / Revised: 24 February 2004 / Accepted: 18 March 2004 / Published online: 20 April 2004
© Springer-Verlag 2004

Abstract Powder metallurgy is a highly developed method of manufacturing reliable ferrous parts. The main processing steps in a powder metallurgical line are pressing and sintering. Sintering can be strongly enhanced by the formation of a liquid phase during the sintering process when using phosphorus as sintering activator. In this work the distribution (effect) of phosphorus was investigated by means of secondary ion mass spectrometry (SIMS) supported by Auger electron spectroscopy (AES) and electron probe micro analysis (EPMA). To verify the influence of the process conditions (phosphorus content, sintering atmosphere, time) on the mechanical properties, additional measurements of the microstructure (pore shape) and of impact energy were performed. Analysis of fracture surfaces was performed by means of scanning electron microscopy (SEM). The concentration of phosphorus differs in the samples from 0 to 1% (w/w). Samples with higher phosphorus concentrations (1% (w/w) and above) are also measurable by EPMA, whereas the distributions of P at technically relevant concentrations and the distribution of possible impurities are only detectable (visible) by means of SIMS. The influence of the sintering time on the phosphorus distribution will be demonstrated. In addition the grain boundary segregation of P was measured by AES at the surface of in-situ broken samples. It will be shown that the distribution of phosphorus depends also on the concentration of carbon in the samples.

Keywords SIMS · AES · PM · Sintering · Activator · Phosphorus

D. Krecar · V. Vassileva · H. Danninger · H. Hutter (✉)
Institute for Chemical Technologies and Analytics,
Vienna University of Technology,
Getreidemarkt 9/164, 1060 Vienna, Austria
e-mail: h.hutter@tuwien.ac.at

Introduction

Powder metallurgy (PM) provides a unique opportunity to produce precision components with complex geometry and excellent surface quality. Most applications of sintered steels are in automotive engineering but non-automotive industry applications are increasing. PM is a suitable method for high volume production of near net shape parts with very high material utilisation. The main disadvantage is the high cost of the powder materials. The properties of PM manufactured products essentially equal those of comparable cast iron, in certain cases these are even superior [1, 2, 3, 4, 5, 6, 7].

For ferrous PM part production certain chemical elements play an important role. These elements, called sintering activators, are able to enhance the sintering processes. They enable attaining of the same mechanical properties at lower sintering temperature and/or shorter sintering time.

Phosphorus, one of these elements, is the focus of this work. Numerous investigations on iron–phosphorus systems have already been performed [1, 2, 3, 4, 5, 6, 7]. Phosphorus addition through Fe_3P powder causes the formation of a transient liquid phase by a eutectic reaction between Fe_3P and Fe during the sintering process at temperatures >1050 °C in a protective atmosphere. The melt is then distributed by capillary forces. This liquid phase provides a better distribution of all other alloying components, a rounding of the pores, and enhances diffusion in α-iron. The maximum solubility of phosphorus in α-iron is 2.55 % (w/w) at eutectic temperature of 1048 °C [8]. Furthermore P stabilizes α-iron with higher self-diffusion of Fe than in γ-iron and leads to higher densities of the sintered parts. Phosphorus addition also increases the hardness of the sintered parts, but at higher contents (above 0.15 % w/w) it decreases the impact energy [9]. Phosphorus is used as a sintering activator in the production of crankshaft sprockets, synchronizer hubs, and synchronizer interlock sleeves [1].

This work describes the phosphorus distribution in PM steels with and without addition of carbon. Influences of

Table 1 Overview of the investigated samples, which contain different concentrations of phosphorus and carbon. All samples were sintered at 1120 °C in a protective atmosphere of flowing hydrogen

Sample	A1	A2	A3	A4	A5	A6	A7	S1	S2	M1	M2
Sintering time (h)	1	1	1	1	1	1	1	1	1	0.5	0.5
P (% w/w)	0	0.15	0.3	0.45	0.6	0.8	1	1	2	0.45	0.6
C (% w/w)	–	–	–	–	–	–	–	1	1	0.7	0.7

the sintering time will be described for samples without carbon content but with fixed a phosphorus amount. It will be shown that techniques like secondary ion mass spectrometry (SIMS) and Auger electron spectroscopy (AES) are very suitable and helpful to supplement the standard analytical techniques used in powder metallurgy (electron probe micro analysis (EPMA) and light microscopy) due to their low detection limits (SIMS) and high lateral resolution (AES).

Experimental

Sample preparation

The investigated sintered samples were produced from water atomized iron powder (ASC 100.29 produced by Höganäs AB, Sweden). The particle size is up to 70% between 45 μm and 145 μm [10]. All samples (except reference sample) contain phosphorus acting as sintering activator (added as ferrophosphorus powder Fe$_3$P, particle size <40 μm). Additionally, samples with different carbon content (by addition of UF 4 nature graphite) were produced to study its influence on the phosphorus distribution.

The iron powder, Fe$_3$P, graphite and HWC (a pressing lubricant) were blended and mixed for an hour in a tumble mixer. Green compacts (55 mm×10 mm×10 mm) were manufactured by uniaxial pressing of the mixture at 600 MPa. Afterwards the green compacts were de-waxed at 600 °C for 30 min and sintered at 1120 °C (1150 °C for M1 and M2) in a pusher furnace for a certain time under a protective atmosphere of flowing hydrogen. Tables 1 and 2 give an overview of the sample composition and the length of the sintering time.

Green densities were calculated from the measured dimensions (length, width, height and mass). Sintering density measurements were performed after resin impregnation by the water-displacement method. Impact toughness measurements were performed by means of Charpy impact tester with W_{Max}=50 J (on unnotched test specimens, ISO 5754). From the microstructural properties only the pore shape will be demonstrated.

Techniques of investigation

All samples were investigated by means of secondary ion mass spectrometry (SIMS). The SIMS device used throughout this investigation was an upgraded Cameca IMS 3 f. The device improvements are mainly in the primary section: an additional primary magnet enables the use of a fine focus Cs$^+$ ion source and a duoplasmatron source (in our case generating O$_2^+$ primary ions); the original beam deflection was replaced by a digital scan generator.

Table 2 Overview of the investigated samples with 0.8% (w/w) P and varied sintering time (length of time in the sintering zone). These samples were also sintered in a protective atmosphere of flowing hydrogen

Sample	P1	P2	P3	P4	P5
Sintering time (min)	5	10	15	20	30

Table 3 Experimental secondary ion mass spectrometry set-up. The primary ion-beam scans rapidly over the sample surface (scanned area) to achieve homogeneous illumination of the sample

O$_2^+$ primary ions	Cs$^+$ primary ions
Beam energy=5.5 keV	Beam energy=14.5 keV
Ip=1.5 μA	Ip=150 nA
Scanned area=300×300 μm^2	Scanned area=300×300 μm^2
Analysed area=150 μm ∅	Analysed area=150 μm ∅
Detected secondary ions= positive	Detected secondary ions= negative

Ip is the primary ion current

ator. In stigmatic mode the Cameca IMS 3f acts as a light microscope whereby each point on the double channel plate corresponds to one certain point on the sample. Both caesium (Cs$^+$) and oxygen (O$_2^+$) primary ions were used to identify the elemental distribution of the sintering activators and other trace elements. Caesium primary ions are used for detecting electronegative elements, e.g. phosphorus, carbon, nitrogen, oxygen, fluorine, sulfur, chlorine and bromine. Oxygen primary ions are used to detect electropositive elements such as, e.g., metals (sodium, aluminium, silicon, potassium, calcium) [11]. Table 3 shows the experimental SIMS conditions for both primary ion beams.

In addition to SIMS measurements, Auger electron spectroscopy (AES, VG Microlab 310F, field emission gun, accelerating voltage: 40 kV), electron probe micro analysis (EPMA, Jeol 6400, accelerating voltage: 20 kV) and light microscopy (to obtain pore shapes and their distribution) measurements were performed to obtain further information and to compare and complement the data with results SIMS. The images of fractured surfaces were recorded by scanning electron microscopy (SEM). AES analyses were performed on the polished surface and on in-situ fractured surfaces (AES spectra and depth profiles).

Results and discussion

Influence of the phosphorus content

As mentioned above, phosphorus as an electronegative element was investigated with caesium primary ions. Nevertheless, measurements with oxygen primary ions were also performed to detect trace metallic components such as alkaline and alkaline earth metals.

For SIMS measurements plane sample surfaces are needed. Thus all samples were ground and polished with diamond paste of varying particle-size. Before SIMS images were recorded, the samples were pre-sputtered for 1 h at maximum primary ion current to obtain really clean surfaces, whereby approximately 1–3 μm of material was removed.

First of all specimen series A (Table 1) was investigated as described above. With increasing phosphorus content, green density is decreases (0.15% (w/w) P, ρ=7.02 g cm^{-3}, 1% (w/w) P, ρ=6.89 g cm^{-3}) because of the increasing

Fig. 1 Microstructure (pore shape) of plain Fe (*left*) and of Fe+1% (w/w) P (*right*). Magnification: 200×; unetched. Increasing the phosphorus content leads to rounding of pores caused by the formation of the liquid phase during the sintering process

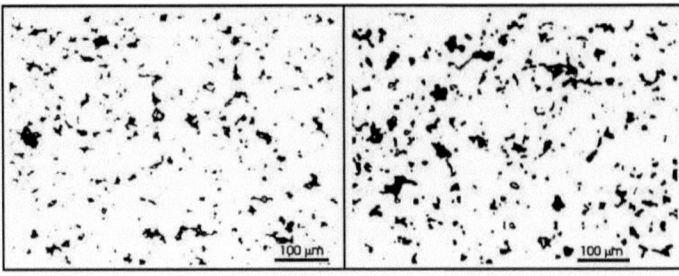

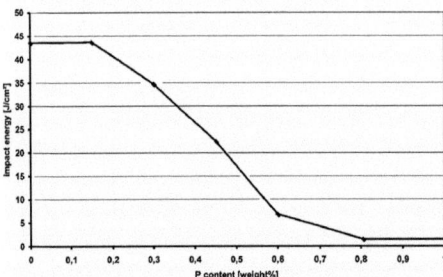

Fig. 2 Impact energy as function of P content. The impact energy decreases significantly above 0.15% (w/w) P

amount of the fine and hard Fe_3P phase (lower compressibility). In contrast to green density, sintered density increases with increasing phosphorus content (0.15% (w/w) P, ρ=7.05 g cm^{-3}, 1% (w/w) P, ρ=7.18 g cm^{-3}). This densification is caused by the enhanced diffusion in α-iron but results also in shrinkage. A further result of liquid phase formation is a spheroidisation of the pore shape. The pores become rounder with increasing phosphorus content (Fig. 1). Impact energy values are slightly increased up to 0.15% (w/w) P and decrease significantly for specimens above 0.15% (w/w) P (Fig. 2).

SIMS investigations (Fig. 3) performed on the samples which were sintered for one hour. They show phosphorus uniformly distributed within all specimens of this series (Fe+x% (w/w) P). Some images (Fig. 3) show brighter points (area), but these are interferences with $^{30}Si^1H^-$ molecular ions. The used powder is technically pure but there is still a silicon content of ≤0.01%. Due to the high sensitivity of the SIMS method for silicon, interferences with silicon are possible. By comparing $^{28}Si^-$ and $^{31}P^-$ images it is possible to distinguish between phosphorus and silicon. These results were consistent with EPMA–EDX mappings of phosphorus. Caesium SIMS measurements also show the presence of oxygen (oxides) and traces of chlorine and carbon. In this series carbon and chlorine are only present as negligible traces in the iron powder. As expected investigations made with oxygen primary ions show the presence of trace elements (metals: Na, Al, K, Ca, Mn, and V), which are detectable at grain boundaries or around pores.

Fractographic analysis of the A series was performed by means of scanning electron microscopy (secondary electron mode). Plain iron (Sample A1) shows ductile fracture; with increasing phosphorus content brittle intergranular fracture occurs (Fig. 4).

The brittle intergranular fracture in Fig. 3 is an indication for the presence of phosphorus at the grain boundaries; otherwise this kind of fracture would not occur. This finding disagrees with previous results achieved by SIMS and EPMA measurements, which did not show P segregation to the grain boundaries [9]. Therefore, Auger electron spectroscopy (AES) measurements were performed, at first on identical samples as used for SIMS of equal composition (ground and polished cross-section) and afterwards on in-situ broken samples. For this purpose cylindrical samples (Ø 5 mm, length 30 mm) with a V notch in the centre of specimens were machined from sintered bars and broken in the high vacuum chamber of an AES device.

AES measurements (Fig. 5) performed on ground and polished sections are in agreement with SIMS results and do not show any phosphorus enrichment at the grain boundaries that could be responsible for brittle fracture as shown in Fig. 4.

AES measurements performed on in-situ broken samples (untreated, fresh broken surfaces) show phosphorus enrichment at the grain boundaries. This enrichment seems to be a monolayer of phosphorus which is responsible for brittle intergranular fracture (Fig. 6). These grain boundary phosphorus enrichments are not observable at metallographic cross sections (Fig. 5) because of the long pretreatment of the samples (grinding and polishing).

Influence of sintering time

Time is a very important factor for phosphorus diffusion into iron grains. All samples shown in series A have been sintered for 1 h. To obtain the sintering time dependence of phosphorus diffusion, samples with 0.8% (w/w) phosphorus were sintered for different lengths of time (Table 2),

Fig. 3 SIMS 2D Images of the phosphorus A series (detected m/e ratio=31). Diameter of images: 150 μm. Primary ions: Cs+. Beam energy: 14.5 keV. Primary ion current: 150 nA. Scanned area: $300\times300\,\mu m^2$

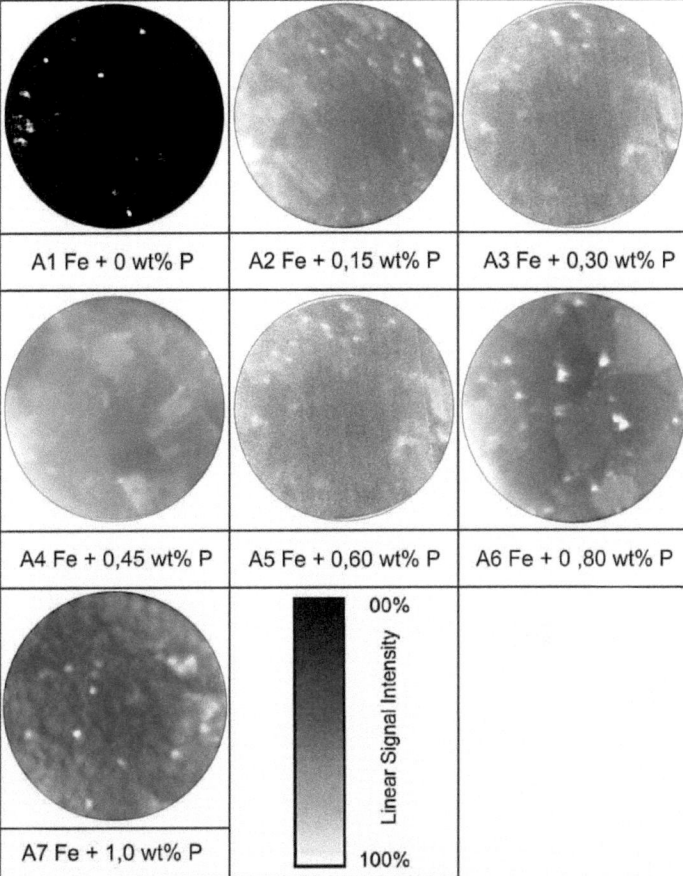

Fig. 4 Scanning electron micrographs of plain Fe (*left*) and of Fe+1% (*w/w*) P (*right*). Plain iron shows ductile fracture in contrast to sample Fe+1% (*w/w*) P where brittle intergranular fracture is observable

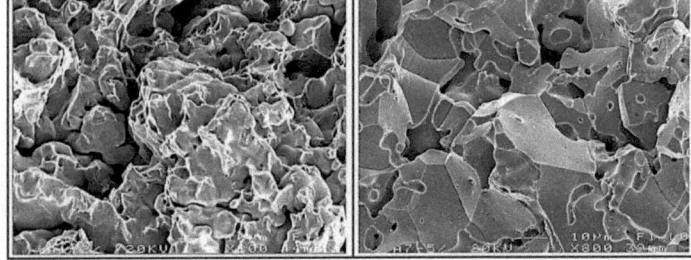

all at a constant temperature of 1120 °C. Figure 7 shows the phosphorus and oxygen distribution after 5 min sintering time. Phosphorus precipitates and enrichments (Fe$_3$P) are observable at the grain boundaries and correlate topologically with oxygen distribution. Obviously the Fe$_3$P particles are still covered by an oxide layer, which inhibits

Fig. 5 SEM images and AES spectra of Fe+1% (w/w) P sample (metallographic cross-section). Black points in *top right* image are measuring points for the AES graphs. AES graphs (the *upper* graph represents the grain and the *lower* the grain boundary) of sample A7. The two AES graphs exhibit no enrichment of phosphorus at grain boundaries

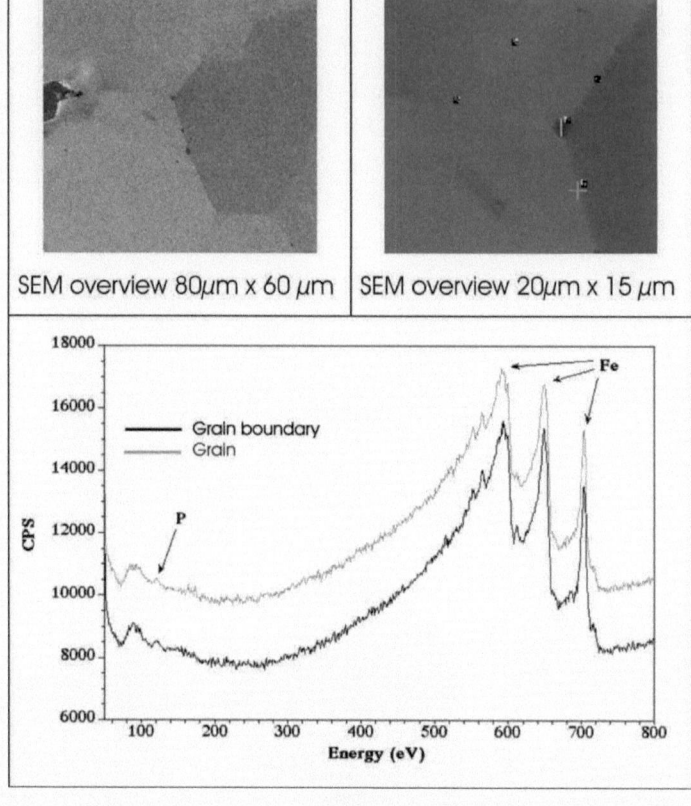

phosphorus diffusion. Five minutes in the sintering zone is too short to destroy the oxide coating.

With increasing sintering time the oxide layer of Fe$_3$P particles is almost completely destroyed and phosphorus diffuses more and more into the iron grains. Figure 8 shows the phosphorus distribution in samples sintered for 10, 20, and 30 min. Figure 8 also shows the oxygen distribution after 10 min at spot 1, where still some oxidized Fe$_3$P particles are available. After 20 min sintering there are still areas observable with lower phosphorus content and finally after 30 min (a completely loss of covering oxide layer) there is almost homogeneous distribution observable as already shown at samples for series A (Fig. 2).

Influence of carbon content

Another component which influences phosphorus diffusion and distribution is carbon. As mentioned above the impact energy decreases rapidly above 0.15% (w/w) P. The main reason of carbon addition is to stop the decrease of impact energy values. Results from initial experiments on carbon distribution and the resulting phosphorus distribution are shown below.

Molinari et al. [12, 13] have already described the mechanism of iron–phosphorus–carbon interaction up to a carbon content of 0.7% (w/w) (Table 4). At sintering temperature austenite and ferrite are present and there is phosphorus enrichment in ferrite. Figure 9 shows a completely different phosphorus distribution of M-samples in comparison to the A-series which does not contain any significant content of carbon.

With higher carbon content austenite formation dominates at sintering temperature (Fig. 10). In the Fe+1% (w/w) P+1% (w/w) C sample phosphorus diffusion into iron grains is inhibited by too high carbon content which stabilizes γ-Fe at sintering temperature. Thereby phosphorus forms a grain boundary network consisting of steadite (γ-Fe+ Fe$_3$C+Fe$_3$P).

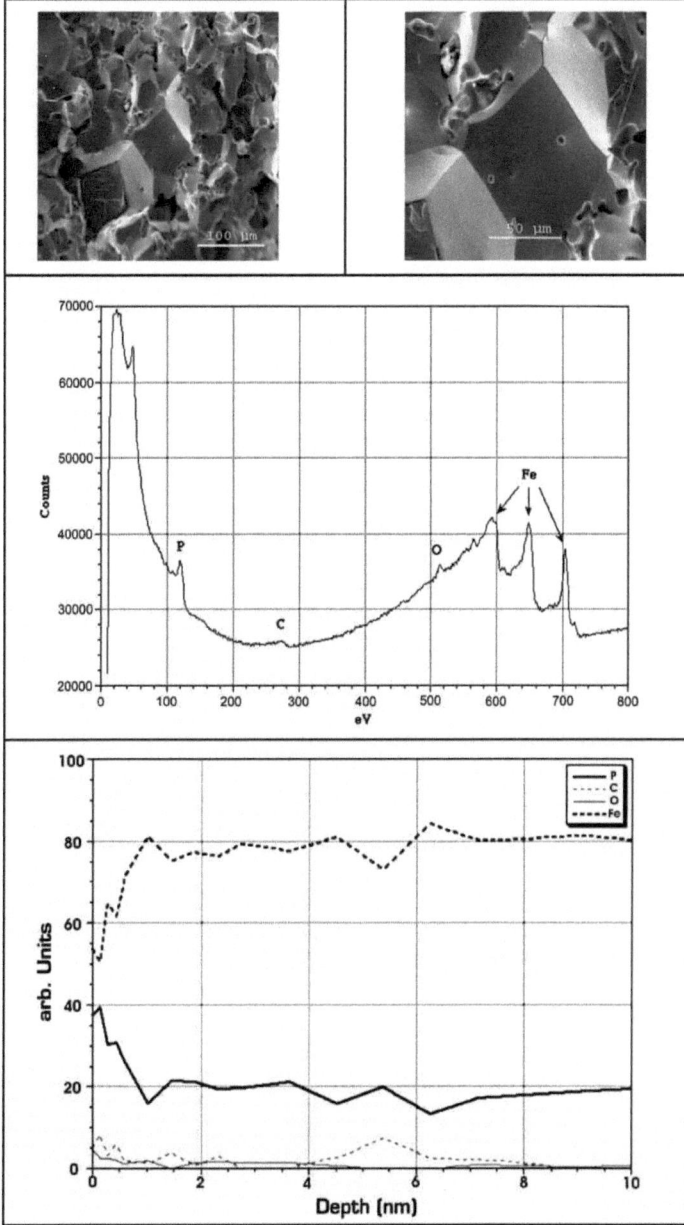

Fig. 6 *Top left*: fracture image (SE image) of in-situ broken sample A6 (Fe+0.8% (w/w) P). *Top right*: fractured area, on which the measurements have been performed. *Centre*: AES spectrum of sample A6 (Fe+0.8% (w/w) P). *Bottom*: AES depth profile performed on the fractured area. Phosphorus enrichment (monolayer coverage) is observable (*full black line*)

Fig. 7 Phosphorus distribution in sample P1 (Fe+0.8% (w/w) P, 5 min) at two different spots and oxygen distribution at spot 1. Sintering time: 5 min. Diameter of images: 150 µm. Primary ions: Cs+. Beam energy: 14.5 keV. Primary ion current: 150 nA. Scanned area: 300×300 µm²

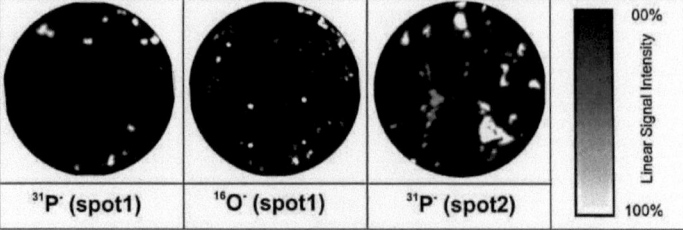

Fig. 8 ³¹P Images in Fe+0.8% (w/w) P sintered for different lengths of time. After 10-min sintering phosphorus diffusion is clearly observable, but nevertheless some precipitates are also present. Detected m/e ratio: 31. Diameter of images: 150 µm. Primary ions: Cs+. Beam energy: 14,5 keV. Primary ion current: 150 nA. Scanned area: 300×300 µm²

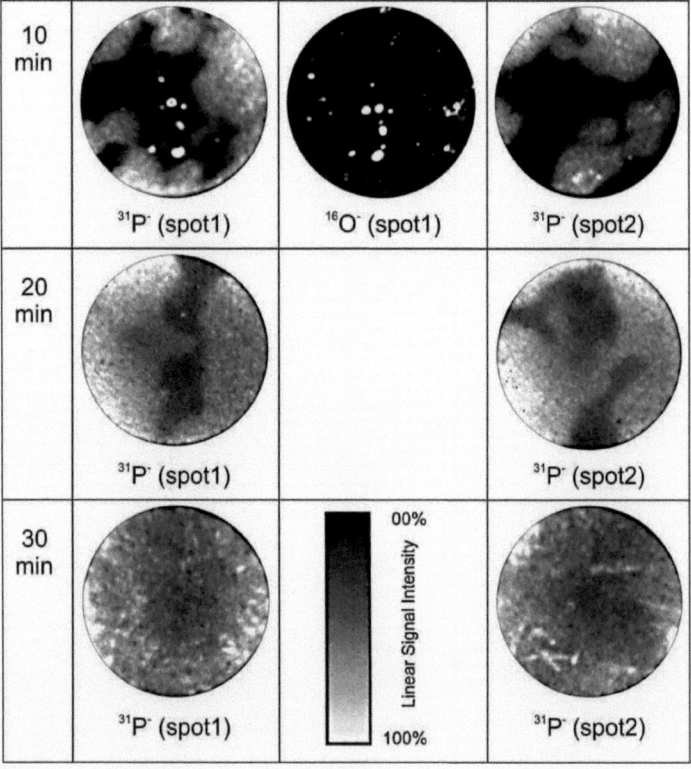

Table 4 Overview of investigated samples containing different amounts of phosphorus and carbon. M samples were sintered for 30 min at 1150 °C under a protective atmosphere of flowing hydrogen. S samples were sintered for 1 h at 1120 °C also under protective atmosphere of flowing hydrogen

Sample	M1	M2	S1	S2
P (% w/w)	0.45	0.6	1	2
C (% w/w)	0.7	0.7	1	1

Conclusions

Phosphorus is a suitable sintering activator for powder metallurgical applications, due to formation of a transient liquid phase by eutectic reactions during the sintering process and by stabilisation of α-iron and resulting enhanced diffusion in α-iron. It causes rounding of the pores, increases the hardness and densification with increasing content, and reduces the impact energy.

Fig. 9 SIMS images of specimens M1 (Fe+0.7% (w/w) C+0.45% (w/w) P) and M2 (Fe+0.7% (w/w) C+0.6% (w/w) P): sintering T=1150 °C, sintering time=30 min, hydrogen. The *bright areas* in carbon maps represent pearlite structures and the *dark areas* ferrite. Phosphorus is predominantly enriched and precipitated in ferrite. Diameter of images: 150 μm. Primary ions: Cs+. Beam energy: 15.5 keV. Primary ion current: 150 nA. Scanned area: 300×300 μm²

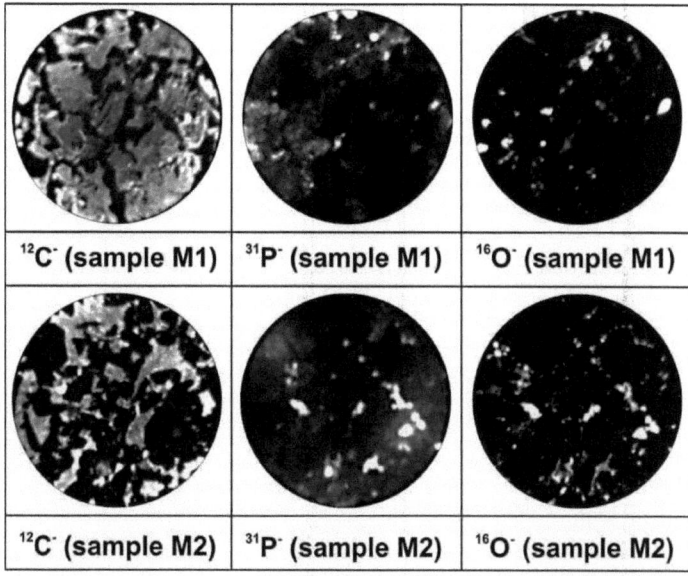

Fig. 10 SIMS images of sample S1 (Fe+1% (w/w) P+1% (w/w) C) with higher carbon content. Carbon is dissolved in iron (austenite), due to much better diffusion behaviour than phosphorus. Phosphorus is enriched in the grain boundary eutectic. Diameter of images: 150 μm. Primary ions: Cs+. Beam energy: 15.5 keV. Primary ion current: 150 nA. Scanned area: 300×300 μm²

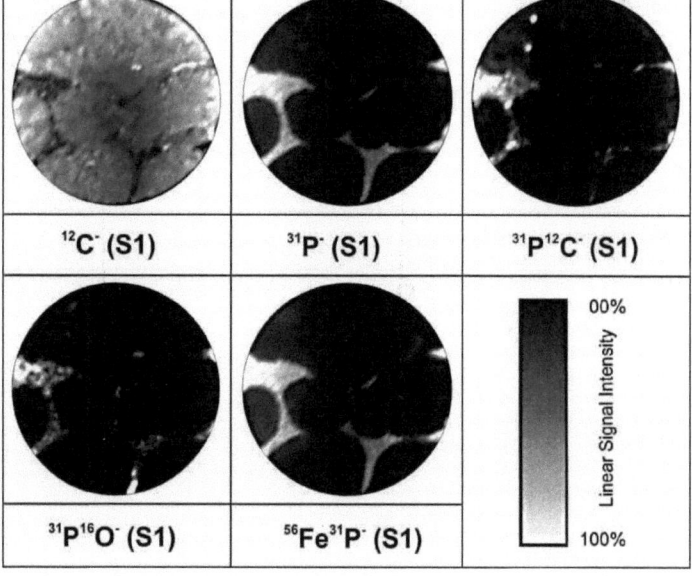

SIMS measurements have shown that if different amounts of phosphorus are added to plain iron, phosphorus is homogeneously distributed after sufficiently long sintering. At higher contents (0.45% (w/w) and above) there is also phosphorus monolayer coverage of the grain boundaries as was demonstrated by AES measurement on in-situ fractured samples. This enrichment is responsible for the observed brittle intergranular fracture. This P segregation could not be shown by analysis of metallographic sections by EPMA, SIMS, and AES.

Measurements of samples sintered for different times have shown that after 5 min almost no diffusion of phosphorus into the iron grains is observable. The ferrophosphorus particles are still covered with oxygen, and thus the formation of a liquid phase and subsequent diffusion are still inhibited. After 10 min most of this oxygen layer is destroyed and phosphorus starts to diffuse. Finally, after 30 min, rather homogeneous distribution is observable.

If carbon is present in the samples up to a content of 0.7% (w/w) C, SIMS measurements show phosphorus enriched in ferrite. At higher carbon content (1% w/w) phosphorus forms a eutectic grain boundary network. Carbon in the eutectic network is better detectable as $m/e=43$ ($^{31}P^{12}C$) than as ^{12}C in SIMS work.

It could be shown that, due to their figures of merit, analytical methods such as secondary ion mass spectrometry and Auger electron spectroscopy are very useful complementary techniques to EPMA for the determination of the elemental distribution of sintering activators and trace elements and impurities.

Acknowledgements The authors want to thank the Austrian Science Fund (FWF) – Project P14889 – for financial support for this work.

References

1. Narasimhan KS (2001) Mater Chem Phys 67:56–65
2. Miura H, Tokunaga Y (1985) Int J Powder Metall Powder Technol 21:269–280
3. German RM, Madan DS (1984) Mod Dev Powder Metall 15:441–454
4. German RM (1998) Powder metallurgy of iron and steel. Wiley, New York
5. Salak A (1995) Ferrous powder metallurgy. Cambridge International Science Publications, Cambridge, UK
6. Thümmler F, Oberacker R (1993) Introduction to powder metallurgy. Institute of Materials, Cambridge University Press
7. Danninger H, Wolfsgruber E, Ratzi R (1977) In: Proc Euro PM '97 on Advances in Structural PM Component Production, Munich, Germany. EPMA, Shrewsbury, p 99
8. Okamoto H (1993) (ed) Phase diagrams of binary iron alloys. ASM, Materials Park, OH, USA
9. Vassileva V, Krecar D, Tomastik C, Hutter H, Danninger H (2003) Effect of phosphorus addition on impact fracture behaviour of sintered iron. Fractography
10. Höganäs (1998) Eisen und Stahlpulver für Sinterformteile. Höganäs, Sweden, pp 79–82
11. Hutter H (2002) Dynamic secondary ion mass spectrometry. In: Bubert H, Jenett H (eds) Surface and thin film analysis, Wiley–VCH, Weinheim, pp 106–121
12. Molinari A, Straffelini G, Fontanari V, Canteri R (1992) Powder Metall 35:285–291
13. Molinari A, Straffelini G, Canteri R (1994) Int J Powder Metall 30:283–291

9.2 Characterization of the Distribution of the Sintering Activator Boron in Powder Metallurgical Steels with SIMS

D. Krecar, V. Vassileva, H. Danninger, H. Hutter

Published in: Analytical and Bioanalytical Chemistry, Springer

Volume 379, Number 4, 2004

605 – 609

SPECIAL ISSUE PAPER

Dragan Krecar · Vassilka Vassileva · Herbert Danninger
Herbert Hutter

Characterization of the distribution of the sintering activator boron in powder metallurgical steels with SIMS

Received: 14 November 2003 / Accepted: 29 January 2004 / Published online: 5 March 2004
© Springer-Verlag 2004

Abstract Powder metallurgy is a well-established method for manufacturing ferrous precision parts. A very important step is sintering, which can be strongly enhanced by the formation of a liquid phase during the sintering process. Boron activates this process by forming such a liquid phase at about 1200 °C. In this work, the sintering of Fe–B was performed under the protective atmospheres of hydrogen, argon or nitrogen. Using different grain sizes of the added ferroboron leads to different formations of pores and to the formation of secondary pores. The effect of boron was investigated by means of Secondary Ion Mass Spectrometry (SIMS) supported by Scanning Electron Microscopy (SEM) and Light Microscopy (LM). To verify the influence of the process parameters on the mechanical properties, the microstructure (pore shape) was examined and impact energy measurements were performed.

The concentrations of B in different samples were varied from 0.03–0.6 weight percent (wt%). Higher boron concentrations are detectable with EPMA, whereas the distributions of boron in the samples with interesting overall concentration in the low wt% range are only detectable by means of SIMS.

This work shows that the distribution of boron strongly depends on its concentration and the sintering atmosphere used. At low concentration (up to 0.1 wt%) there are boride precipitations; at higher concentration there is a eutectic iron–boron grain boundary network. There is a decrease of the impact energy observed that correlates with the amount of eutectic phase.

Keywords SIMS · Sintering · Activator · Boron · PM

D. Krecar · V. Vassileva · H. Danninger · H. Hutter (✉)
Institute for Chemical Technologies and Analytics,
Vienna University of Technology,
Getreidemarkt 9/164, 1060 Vienna, Austria
Tel.: +43-158801-15120
e-mail: h.hutter@tuwien.ac.at

Introduction

The production of industrial precision parts by powder metallurgy is a well-proven technique. Theere are many applications of this process in the automotive industry, including the production of engine parts like crankshaft sprockets or belt drives, but also non-automotive applications of the process are also increasing [1, 2, 3].

For industrial production, the use of sintering activators such as phosphorus or boron is possible [4]. These elements can enhance sintering by forming a liquid phase during the isothermal sintering process; in case of boron by the eutectic reaction of iron and ferroboron at $T \geq 1170$ °C. In general, the liquid phase formation enables a higher degree of densification, rounding of the pores, sintering at lower temperatures and for shorter times, meaning that mechanical properties of the parts are improved.

The quality of sintered parts depends on the distribution of the main components, the alloying elements, and also of trace elements and impurities. Because of its low detection limit (sub-ppm), Secondary Ion Mass Spectrometry – SIMS (which yields 2-D and 3-D elemental distributions of the desired components) – can be a very useful method for examining the elemental distribution of boron, trace elements and impurities in a metal matrix. In this work, the distribution of mainly boron, as the sintering activator, was investigated by means of SIMS. The boron distribution in investigated samples strongly depends on the boron content and on the sintering atmosphere.

Experimental

Sample preparation

The investigated sintered samples were produced from water atomized iron powder (ASC 100.29, with 70% of particle sizes in the range from 45–145 µm, produced by Höganäs AB, Sweden). Ferroboron was added to all samples, which acted as a sintering activator (Fe18B, <40 µm).

The iron powder, ferroboron and pressing lubricant (HWC) were blended for an hour in a tumble mixer. Green compacts (55 mm×

Table 1 Overview of the samples investigated with different concentrations of boron

BH: samples sintered under a protective atmosphere of flowing hydrogen; BA: argon protective atmosphere; BN: nitrogen protective atmosphere

Protective atmosphere	Boron content (wt%)						
	0.03	0.06	0.1	0.015	0.2	0.3	0.6
Hydrogen	BH1	BH2	BH3	BH4	BH5	BH6	BH7
Argon	BA1	BA2	BA3	BA4	BA5	BA6	BA7
Nitrogen	BN1	BN2	BN3	BN4	BN5	BN6	BN7

Table 2 Parameters for experimental SIMS setup

Parameter	Value
Primary ions	Cs^+, O_2^+
Secondary ions	negative (@Cs^+), positive (@O_2^+)
Beam intensity	150 nA (@Cs^+), 1 µA (@O_2^+)
Beam energy	14.5 keV (@Cs^+), 5.5 keV (@O_2^+)
Scanned area	300×300 µm^2
2-D image diameter	150 µm

10 mm×~10 mm) were produced by uniaxial pressing of the mixture at 600 MPa. The dimensions and the mass of the specimen bars were measured, and green densities were calculated. Then the green compacts were de-waxed at 600 °C for 30 min and sintered at 1200 °C in a pusher furnace for 1 h. Sintering was performed in three different protective atmospheres: hydrogen, nitrogen (5.0), and argon (5.0), to obtain the dependence of the protective atmosphere on the sintering process. Table 1 gives an overview of the investigated samples.

Sintering density measurements were performed by the water displacement method. The impact energy was determined using a Charpy impact tester (W_{max}=300 J).

Investigation techniques

The main investigation technique used throughout this work was Secondary Ions Mass Spectrometry – SIMS. The measurements were performed with an upgraded Cameca IMS 3f. The improvements made to the device are mainly in the primary section. Both cesium (Cs^+) and oxygen (O_2^+) primary ions are used to identify the elemental distribution of the elements of interest (sintering activator, trace elements or impurities). Cesium primary ions are used to detect electronegative elements such as phosphorus, carbon, nitrogen, oxygen, chlorine, and sulfur, as well as oxides and nitrides. Oxygen primary ions are used to detect less electronegative elements such as the metals sodium, aluminum, chromium, vanadium, and manganese, as well as boron [5].

In stigmatic mode, the Cameca IMS 3f acts like a light microscope, whereby each point on the sample surface corresponds to a point (channel) of the double channel plate. By means of a CCD camera, it is possible to record 2-D images of the elemental distributions. For 3-D imaging, 64 images of each desired mass are recorded during the sputtering process, down to a typical depth of 5 µm. These 64 images are compiled into one 3-D cube using a program, "Visualizer 2" [6], based on the vtk (visualisation tool kit) library, and developed in-house. Table 2 shows the experimental SIMS setup.

In addition to SIMS measurements, fractographic analysis was performed using scanning electron microscopy (SEM), and the pore shape was observed using a light microscope.

Results and discussion

SIMS measurements need plane sample surfaces, and therefore all samples were ground and polished with diamond

Fig. 1 Light microscope image of sample (BA7, Fe+0.6 wt% B) surface after primary ion treatment. Primary ion beam acts as an etchant: the network of grain boundaries as well as the pores are clearly observable

paste. Before 2-D images were recorded, the samples were pre-sputtered ("cleaned") by means of the primary ions for 30 min. The primary ion beam acts as etchant. Figure 1 shows the cross-section of the sample BA7 after the primary ion treatment. In this way, approximately 1–3 µm of material was removed.

Sintering in hydrogen and argon atmosphere

The addition of boron leads to the formation of a persistent liquid phase produced by the eutectic reaction of iron and ferroboron at temperature above 1170 °C. This liquid phase causes rounding of the pores (Fig. 1) and increases the sintered density as shown in Fig. 2. The green density decreases with increasing boron content because of the lower density and compressibility of the hard ferroboron powder. There is also an increase in observed impact energy values up to a boron content of 0.1 wt%. Above this concentration the values decrease.

The use of coarse ferrroboron powder (Fe21B, >40 µm), which melt during the sintering process, causes the formation of larger secondary pores (Fig. 3). Sintering performed under a protective atmosphere of flowing hydrogen leads to a reaction between boron and hydrogen at the edges of the specimens (forming gaseous B_2H_6), which results in a slight depletion of boron in these areas (Fig. 3) [7].

Fig. 2 Comparison of green and sintering density (protective atmosphere: hydrogen). The impact energy values are also shown (y-axis on the right hand side of the plot)

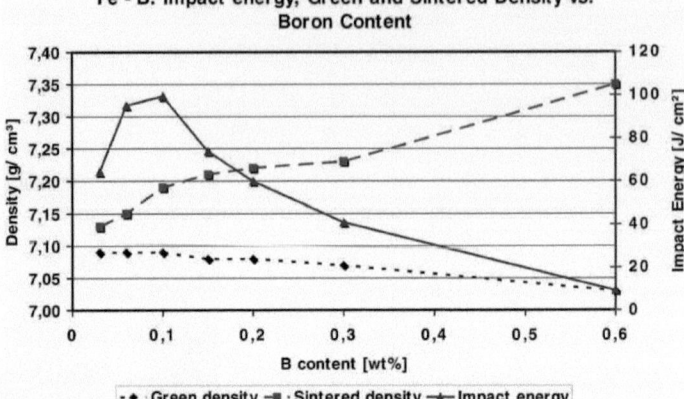

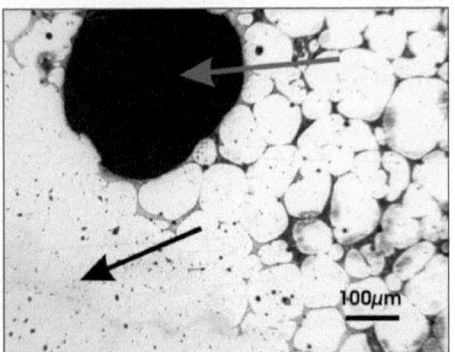

Fig. 3 Cross-section of Fe+0.6 wt% B (sintering atmosphere: hydrogen). The black arrow shows the area of the boron depletion (edge of specimen); the gray arrow points to a secondary pore

Fractographic analysis of the Fe+x wt% B – samples sintered under hydrogen atmosphere show ductile fracture, with some transgranular cleavage areas at lower boron content (up to 0.1 wt% B) and transgranular fracture with some plastic deformations at higher boron content (see Fig. 4).

SIMS measurements were mainly focused on the recording of 2-D (lateral elemental distribution) and 3-D images of boron, but trace elements and impurities were also investigated. Oxygen primary ions were used to identify the distributions of boron and metallic components. If sintering is performed in a protective atmosphere of flowing hydrogen, boron precipitation at grain boundaries is observable. At higher concentrations (0.15 wt% and above), besides boron precipitation a grain boundary network of boron and iron is formed. During the sintering process, a liquid phase of Fe and Fe_2B is formed by eutectic reactions. Because of the very low solubility of boron in iron, a grain boundary network of iron and boron remains (Figs. 5 and 6).

These results correspond well with the impact energy values already shown in Fig. 2. A rapid decrease in impact energy values is accompanied by the formation of a grain boundary network.

Further SIMS investigations of these samples, performed with both oxygen and caesium primary ions, show traces of sodium, aluminum, silicon, chromium manganese,

Fig. 4 Scanning electron micrographs of Fe+0.06 wt% B (left) and Fe+0.6 wt% B

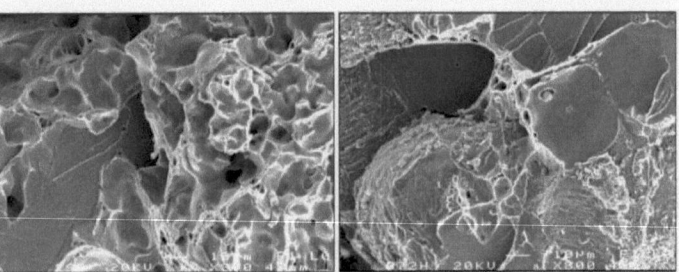

Fig. 5 SIMS 2-D images of a boron concentration series sintered in a protective atmosphere of flowing hydrogen. Diameter of images: 150 µm

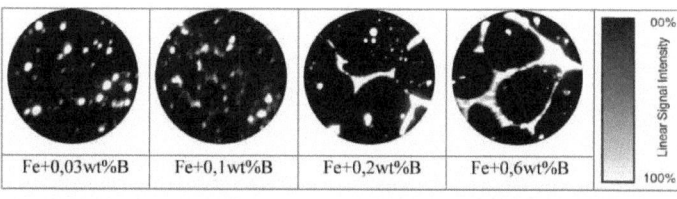

Fig. 6 3-D SIMS images of boron distribution in isosurface mode. In isosurface mode, points with the same intensity are displayed as one surface. Dimensions: lateral 150 µm × 150 µm, depth 5 µm

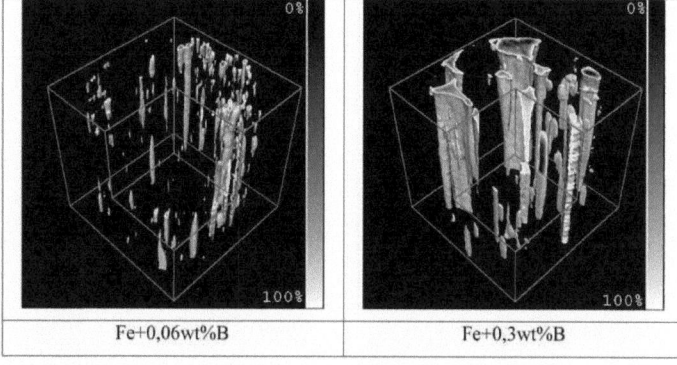

Fig. 7 SIMS 2-D images of a boron series with increasing boron content, sintered in a protective atmosphere of flowing nitrogen. Diameter of images: 150 µm

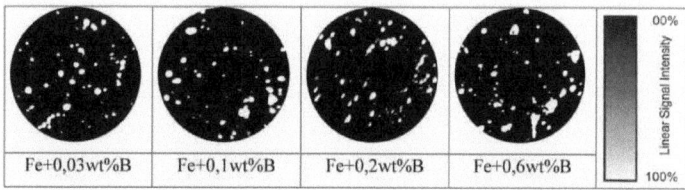

Fig. 8 3-D SIMS images of boron distribution in isosurface mode for two different samples sintered in a protective atmosphere of nitrogen. Dimensions: lateral 150 µm × 150 µm, depth 5 µm

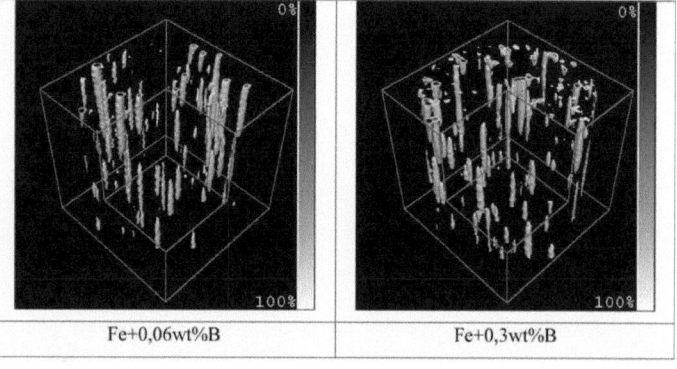

Fig. 9 SIMS 2-D images of different elements and fragments (oxides, nitrides and oxynitrides) of an Fe+0.3 wt% B sample sintered in a protective atmosphere of nitrogen. Diameter of images: 150 μm

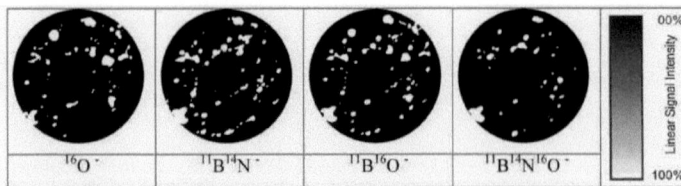

oxygen, chlorine, sulfur, and oxides, originating from the powders used. These impurities are concentrated at the grain boundaries or surrounding pores.

Using argon as a protective sintering atmosphere leads to similar results (impact energy, fractographic analysis and boron distribution) to those described above for samples sintered in a hydrogen atmosphere. The main difference is the inertness of argon compared to hydrogen. Argon is a less reactive gas, so no reaction between argon and boron occurs; in particular, no boron depletion at the edge of the specimen is observed. The drawback of argon usage is the very poor removability of argon, which fills the pores during sintering and stabilizes closed pores against shrinkage, inhibiting densification.

Sintering in nitrogen atmosphere

If a nitrogen protective atmosphere is used, completely different results are obtained. The boron distribution is the same throughout the whole series. Only precipitations of boron are observable (Figs. 7 and 8), no eutectic network was formed, even at high boron amounts. The reason for this behavior is the high affinity of boron to nitrogen [8]. The formation of different nitrides and oxynitrides during the sintering process is possible (Fig. 9). The presence of nitrogen inhibits eutectic reactions between iron and iron boride, but enhances the reactions between boron and nitrogen, resulting in the formation of an inert phase that does not improve the sintering rate (on the other hand, BN is used for the activation of sintering in vacuum [9]). There was therefore no improvement in measured mechanical properties.

Conclusions

Boron is a useful sintering activator if sintering is performed under a protective atmosphere of hydrogen or argon. Due to the eutectic reaction between Fe and Fe_2B, a persistent liquid phase is formed during the sintering process, which provides densification and pore rounding. An improvement in the impact energy values is observed up to 0.15 wt% B. These values correspond to SIMS measurements, where boron is precipitated at the grain boundaries up to 0.15 wt%. Due to the very low solubility of boron in iron, boron forms a grain boundary network in samples above 0.15 wt%. During the sintering process, a reaction of boron with hydrogen occurs to form B_2H_6, which leads to a slight depletion of boron at the edge of specimens.

If sintering is performed in argon, similar results are obtained as with a hydrogen protective atmosphere. Because of the lower reactivity of argon, there is no boron depletion at the edge of samples. However, using an argon atmosphere is more expensive than using hydrogen, and it is not possible to remove argon from closed pores during sintering, which adversely effects densification. This could be a drawback if further treatment of sintered parts is planned.

Nitrogen is not an acceptable sintering atmosphere, due to its high affinity to boron. No eutectic reactions between Fe and Fe_2B occur, but reactions do occur between boron and nitrogen. There is also no improvement in the measured mechanical properties in this case.

To sum up, Secondary Ion Mass Spectrometry is very useful technique for 2-D and 3-D determination of elemental distribution of sintering activators, as well as of trace elements and impurities in PM steels.

Acknowledgements The authors want to thank the Austrian Science Fund (FWF) – Project P14889 – for the financial support of this work.

References

1. Narasimhan KS (2001) Mater Chem Phys 67:56–65
2. Salak A (1995) Ferrous powder metallurgy. Cambridge Int Science, Cambridge, UK
3. German RM (1998) Powder metallurgy of iron and steel. Wiley, Chichester, UK
4. Benesovsky F, Hotop W, Frehn F (1955) Planseeber Pulvermet 3:57
5. Hutter H (2002) Dynamic secondary ion mass spectrometry. In: Bubert H, Jenett H (eds) Surface and thin film analysis. Wiley-VCH, Weinheim, Germany, pp 106–121
6. Hutter H, Nowikow K, Gammer K (2001) Appl Surf Sci 179: 161–166
7. Selecka M, Dudrova E, Bures R, Kabatova M, Salak A (1995) Pokroky Praskove Metalurgie 1–2:77–86
8. Molinari A et al. (1994) Powder Metall 37(2):115–122
9. Danninger H, Jangg G, Giahni M (1988) Materialwiss Werkst 19:205–211

9.3 SIMS Investigation of Cr – Mo Low Alloyed Steels Sintered With Boron

D. Krecar, V. Vassileva, H. Danninger, H. Hutter

Published in: Submitted to Analytical and Bioanalytical Chemistry, Springer

SIMS Investigation of Cr Mo Low Alloyed Steels Sintered With Boron

D. KRECAR, J. ZWANZIGER, V. VASSILEVA, H. DANNINGER, H. HUTTER*
Institute of Chemical Technologies and Analytics, Vienna University of Technology
Getreidemarkt 9 / 164 AC, A 1060 Vienna
* For correspondence: Email: h.hutter@tuwien.ac.at Phone: +43 1 58801 15120

Keywords: SIMS, sintering, activator, boron, PM,
PACS: 68.49.Sf., 82.80.Ms., 81.20.Ev.,

Abstract

The sintering activator boron enhances the sintering process by the formation of the liquid phase during the sintering process by the eutectic reaction of iron and ferroboron at a temperature above 1170°C. The effect of boron was investigated by means of 2D and 3D secondary ion mass spectrometry (SIMS) supported by scanning electron microscopy (SEM). Hereby the samples were sintered for 60 min with boron amounts in the range of 0 0.6 wt% using two different chromium (1.5 wt% and 3 wt%)- and molybdenum (0.2 wt% and 0.5 wt%)- pre-alloyed powders at two different temperatures (1200° and 1300°C) and under two different sintering atmospheres (hydrogen and argon). Additionally same samples with an significant amount of carbon (0.6 wt%) were also investigated. The influence all of these parameters on the boron distribution which differs from the strongly depletion at the specimen surface, irregular precipitate formation at lower contents, boron inclusions inside the steel matrix and the formation of the grain boundary networks at the core of the samples will be discussed. The influence of boron on the physical and mechanical properties on the produced materials (steels) is evaluated additionally by measurements of hardness, impact toughness, density and fractography.

Introduction

Nowadays the materials production depends primarily on the desired application requirements of these materials but the resulting production costs plays also an important role. For many years the powder metallurgy became an alternative method for low cost mass production of the precision components with complex and excellent geometry and with very high material utilization mostly in automotive- but also in non-automotive industry [1, 2 ,3, 4]. For the part production the sintering process can be improved in some ways. One of these enhancements can be reached using sintering activators, whereby the motivation is shortening the sintering time and lowering the sintering temperature and production costs but still maintaining the same or even improving the desired material properties as without the usage of these activators. The sintering activator (e.g. boron, phosphorus, copper) are causing the formation of the liquid phase during the sintering process. Boron is known as one of the very good sintering activators for sintering with ferrous systems. In case of sintering with boron the permanent liquid phase is formed during the isothermal sintering process by the eutectic reactions of ferroboron with iron at the temperatures above 1170°C. In principle the formation of the liquid phase enables the better distribution of all alloying elements, the strong densification of the material and rounding of the pores. Nevertheless there are also some drawbacks using boron as sintering activator: chemical reaction with some sintering atmospheres (e.g. nitrogen), eutectic grain boundary network formation, embrittlement, geometry problems of specimens caused by the liquid phase formation and to high densification and secondary pores. [1, 5, 6, 7, 8, 9].

The preliminary investigation was concentrated only on plain Fe-B samples sintered at 1200 °C under protective atmospheres of argon, hydrogen and nitrogen [7]. In this work four different parameters on sintering with boron will be discussed and presented. In order to increase the mechanical properties the influence of the boron, carbon, chromium and molybdenum concentration was studied and two different sintering temperatures were used. [9,]. Due to the distribution of the sintering activator as well as of alloying and trace elements have a big influence on the quality of the obtained parts, 2D- and 3D- secondary ion mass spectrometry measurements are promised to be helpful for this kind of investigations and were applied to the sintered samples. The mechanical and physical properties of the obtained material will be described on the basis of the impact toughness, fractography, density and hardness measurements.

Experimental

Sample preparation
The prealloyed powders, ferroboron and pressing lubricant (HWC) were blended for 60 min in a tumble mixer. The compaction of the green compacts (55mm x 10mm x 10mm) was made by uniaxial pressing at 600 MPa. After de-waxing at 600°C for 30 min the green compacts were sintered for 60 min at 1200°C respectively at 1300°C in a pusher furnace. If carbon (0.5 wt%, UF 4 nature graphite) was not added to the starting mixture, thus a reducing hydrogen sintering atmosphere was chosen, otherwise argon was used.

The green densities were calculated from the measured specimen bars dimensions (length, width, height) and mass. Sintering density measurements were made after resin impregnation by means of water displacement method. Table 1 gives the summary of the investigated samples.

Techniques of investigation
The major part of the investigations in this work was made by means of secondary ion mass spectrometry (SIMS). The SIMS instrument used throughout this investigation was an upgraded Cameca IMS 3f. In stigmatic mode Cameca IMS 3f acts as an ion microscope whereby each point on the double channel plate corresponds to one certain point on the sample surface.

Using a CCD camera, it is possible to record 2D images of elemental distributions. For 3D presentation, 64 images of investigated masses are recorded during the permanently sputtering process, down to a typically depth of 4 5 μm, and compiled into one 3D cube using an in house developed software "Visualizer2" (based on free vtk visualisation tool kit library) [11].

Both, caesium and oxygen, primary ions were used for the generation of subsequently detected secondary ions. Oxygen ions were used to determine the elemental distribution of the main components of the investigations: boron, chromium, iron and molybdenum, but also of some impurities and trace elements. Caesium ions were mainly used for the determination of the elemental distribution of carbon and oxygen [12]. All SIMS measurements were performed on plane metallographic cross sections. Therefore the samples were ground and polished with diamond paste. Before recording 2D element distribution images, the samples were pre-treated with the primary ion beam in order to reach a clean surface. Thereby approximately 2 3μm of material were removed. The experimental SIMS setup is shown in the table 2.

In addition to SIMS, the fractographic analysis was performed using scanning electron microscopy (SEM) in secondary electron mode. Apparent hardness measurements (HV10) and the impact energy measurements, by means of Charpy impact tester with WMax = 300 J (on unnotched test specimens, ISO 5754) were also performed to get the information on material properties after the sintering process.

Table 1: Overview of the investigated samples. Additionally, sintering temperature and the sintering atmosphere are given. (* The investigated B concentration steps were:0, 0.03, 0.06, 0.1, 0.15, 0,2, 0.3 and 0.6 wt%)

Sample	Powder	Cr [wt%]	Mo [wt%]	C [wt%]	B* [wt%]	Sintering T [°C]	Sintering Atmosphere
BCrL1200	AstaloyCrL	1.5	0.2	--	0 – 0.6	1200	H_2
BCrM1200	AstaloyCrM	3.0	0.5	--	0 – 0.6	1200	H_2
BCrLC1200	AstaloyCrL	1.5	0.2	0.6	0 – 0.6	1200	Ar
BCrMC1200	AstaloyCrM	3.0	0.5	0.6	0 – 0.6	1200	Ar
BCrL1300	AstaloyCrL	1.5	0.2	--	0 – 0.6	1300	H_2
BCrM1300	AstaloyCrM	3.0	0.5	--	0 – 0.6	1300	H_2
BCrLC1300	AstaloyCrL	1.5	0.2	0.6	0 – 0.6	1300	Ar
BCrMC1300	AstaloyCrM	3.0	0.5	0.6	0 – 0.6	1300	Ar

Table 2: Experimental Secondary Ion Mass Spectrometry setup. The primary ion beam scans rapidly over the sample surface (scanned area) to achieve a homogeneous illumination of the sample.

	O_2^+ primary ions	Cs^+ primary ions
beam energy	5.5 keV	14.5 keV
primary ion current	500 nA	150 nA
scanned area	250 x 250 µm²	250 x 250 µm²
analyzed area (diameter)	150 µm	150 µm
detected secondary ions	positive	negative

Result and discussion

In this work the influence of two different used pre-alloyed powders respectively the influence of chromium and molybdenum, sintering temperature, significant amount of carbon and different boron contents on the sintering process and on the mechanical properties of the sintered parts will be discussed. The water atomised powders used throughout this work were prealloyed with 3 wt% Cr + 0.5 wt% Mo (samples marked with BCrM) respectively with 1.5 wt% Cr + 0.2 wt% Mo (samples marked with BCrL).

Figure 1 and 2 show that the addition of the well defined increasing amount of hard ferroboron, which has a worse compressibility and lower density than basic powders, results in a almost constant decrease of the green density. The significant addition of one other low destiny component, carbon, leads also to a further slightly decrease of the green density (Figure 1 and 2). During the sintering process the persistent liquid phase is formed caused by the originally addition of the ferroboron. The liquid phase formation is the result of the eutectic reaction between iron and ferroboron at the temperature about 1170°C [5, 6, 7]. The formation of this liquid phase causes the rounding of the pores and it results at first at the densification of the material if no significant amount of carbon is added. Figure 1 shows also that the increase of the sintering temperature from 1200°C too1300°C, pronounce a stronger densification and leads to an enhancement of the sintering density values due to higher diffusion of atoms and this results in stronger sintering necks.The less amounts of the alloying elements chromium and molybdenum (samples marked with BCrL) have a marginal influence on the green and sintering density (no figure), i.e. BCrM and BCrL behave here very simmilary. If sintering is performed with 0.6 wt% carbon the sintering density is at first increased up to a

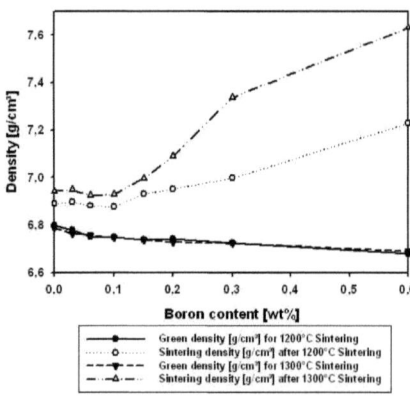

Figure 1: Green and sintering densities for BCrM (H_2 sintering atmosphere) samples depending on sintering temperature and boron content.

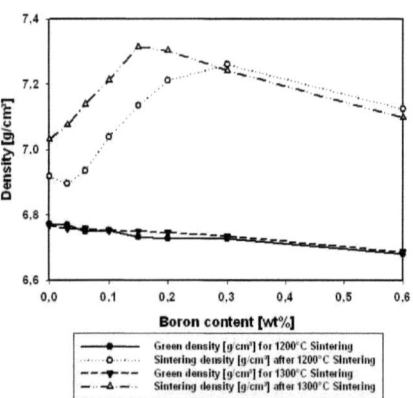

Figure 2: Green and sintering densities for BCrMC (0.6 wt% C, Ar sintering atmosphere) samples depending on sintering temperature and boron content.

boron content of 0.3 wt% (at 1200°C) respectively 0.15 wt% (at 1300°C) (figure 2).Above these two values (higher boron contents) a decrease of the sintering density was found. The bigger the boron additions the higher the amount of the liquid phase. When carbon is alloyed a reduction of iron oxides (from the used prealloyed powder particles) takes place and CO is generated. If the liquid phase has densified the material and the gas has no possibility to leave via pores, CO was enclosed. On the sample surface bubbles were found and the cross sections shows additional pores the sintering density decreases.

It is well known, that boron enhances the sintering process due to formation of the liquid phase during the sintering process. The resulting boron distribution in the investigated samples depends primarily on its content in the samples but also on the presence respectively on the absence of carbon in the samples. If sintering is performed without a significant amount of carbon there is different boron distribution in the core of the test specimens then at the surface of the investigated cross sections. The sintering atmosphere is hydrogen and boron was removed by hydrogen, especially at the surface. Since boron is randomly distributed in form of eutectic precipitates up to an amount of 0.3 wt%, above this content it forms the grain boundary network as shown in figure 3. The formation of grain boundary networks leads to the densification of the samples and to higher sintering density values as shown in figure 1. Figure 3 (lower row) shows that even higher boron amounts do not lead to a formation of grain boundary network on the surface, but it can but it can be observed in the core of the samples. There is nearly always a gradient in the boron distribution and concentration throughout the specimens without a significant amount of carbon and with one nominal boron concentration The previous works [13, 14] on this matter do not mention this effect of boron depletion. Here, this phenomenon of the boron depletion is demonstrated in figure 4 by means of SIMS-, light microscopy- and SEM-images. A line scan from the edge to the inside of sample shows at first a nearly completely depletion at the edge, randomly distributed boron precipitations at the range of 400 – 500 µm towards inside the sample and at last the grain boundary network formation. Additionally the presence of the secondary pores is also demonstrated at the light microscopy- and SEM-image.

Figure 5 shows the impact energy values depending on boron content, carbon content and sintering temperature. In general for samples sintered without carbon content the impact energy is increased up to a boron content of 0.2 wt% and then decreases rapidly. The reason of the very irregular curve progression of the series BCrM (sintered at 1200°C) is probably the irregular (core and surface) and inhomogeneous distribution of boron throughout the test bars as demonstrated above in figures 3 and 4.

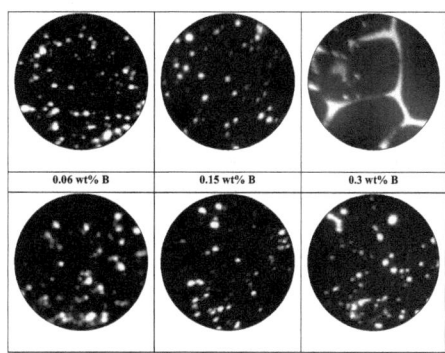

Figure 3: Comparison of the boron distribution for BCrM samples obtained in the middle (upper row) and at the edge of the investigated cross section. Boron depletion occurs at the sintering process, best viewed at the sample with 0.3 wt% B where no eutectic grain boundary network is formed (lower row). Diameter of the images: 150µm, linear signal intensity: white:100%, black:0%.

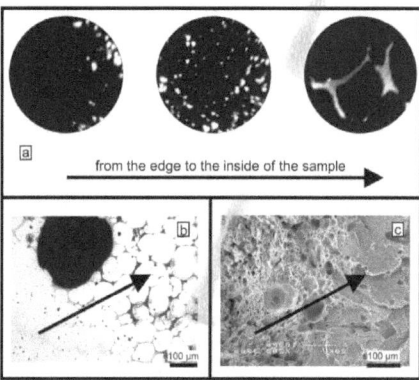

Figure 4: a) SIMS 2D images of the boron distribution during a line scan as indicated with the arrow at the light microscopy image (b) and at the secondary electron image (c). The sample was sintered with 0.6 wt% B, 3 wt% Cr and 0.5 wt% Mo under sintering atmosphere of flowing hydrogen at 1200°C. Diameter of the SIMS images: 150µm, linear signal intensity: white:100%, black:0%.

The increase of the impact energy at the start is mainly caused by the fortification of the sintering necks and the enhancement of the α iron self diffusion, supported by the formation of the liquid phase. With increasing boron amount, 0.3 wt% and above, the eutectic formed grain boundary networks, which are also characterised by means of SIMS above, and additional formation of the secondary pores provide the worse impact energy values. The higher sintering temperature leads here to the nearly same curve progression, but the impact energy values are higher due to temperature induced enhancement of the sintering necks and of the α iron self diffusion.

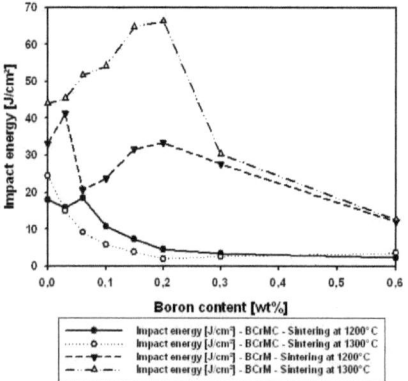

Figure 5: Impact energy behaviour for samples sintered with and without a significant carbon amount depending on sintering temperature and boron content.

Sintering with an significant amount of carbon (0.6 wt%) occurs in the protective atmosphere of flowing argon and so decarbonising caused by hydrogen can not take place. In contrast to above showed samples, if sintering is performed with carbon and argon as sintering atmosphere, there is no boron removal at the surface of the investigated specimens observable. Carbon influences the sintering process, due to its rapid diffusion into the metal matrix; it seems that carbon constrains boron and the formation of an eutectic grain boundary network at lower boron amounts (0.15 wt%) then without carbon (0.3 wt%) was found (figure 6, lower row). Besides the boron precipitation, at lower boron content, as presented till now, figure 6 shows also boron distributed inside the iron grains. At presence of chromium and molybdenum, boron diffuses, due to its high affinity to those elements, in the matrix and a reaction with the matrix is possible [9]. The second reason of boron presence in the matrix is that boron particles are situated from beginning on on these places and do not react or form a liquid phase with iron. During the sintering process the iron grains are growing and enclose these originally particles. The micro hardness of the eutectic phase and of these enclosed particles is the same.

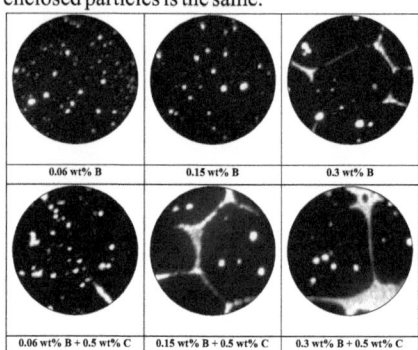

Figure 6: Comparison of the boron distribution for samples BCrL (uper row) and BCrLC (lower row). The formation of the grain boundary network starts already at 0.15 wt% if sintering is performed with the significant amount of carbon. Diameter of the images: 150µm, linear signal intensity: white: 100%, black: 0%.

The addition of carbon leads to a rapidly decrease of impact energy values, already at low boron contents (figure 5). The increasing sintering temperature or the different amounts on alloying elements chromium and molybdenum show rather no difference in the impact energy behaviour of the investigated samples. The addition of 0.6 wt% carbon leads to occurrence of the bainitic structure, which leads on one hand to higher hardness values but on the other hand to an embrittlement of the material (figure 9) and to reduced bulk ductility.
At the presence of carbon, the grain boundary network formation starts already at 0.15 wt% boron which here already badly influences the impact energy values. This negative influence can also be shown by fractography analysis. Figure 7, scanning electron micrographs in secondary electron mode, shows a typically ductile fracture for the sample with the lower boron amount (0.06 wt%) and a transgranular cleavage fracture for the sample with higher boron amount (0.6 wt%) which is evident for the embrittlement of the material. For apparent hardness of chromium and molybdenum play an important role. The higher amount of chromium and molybdenum leads to higher hardness values due to the formation of different hard chromium- and molybdenumborides or even carbides, which is enhanced at higher amounts

of the involved elements. Figure 8 shows typically 3D SIMS images of the main components (B, Cr and Mo) of investigated samples, whereby a very well correspondence of the in depth distributions of these three elements is observable. The grain boundary network of boron as well as above mentioned boron distribution inside the grains is evident. The images demonstrates the homogenously distribution of chromium and molybdenum inside the iron matrix but also the formation of the chromium and molybdenum borides or even chromium molybdenum mixed borides, which are primarily responsible for higher hardness values.

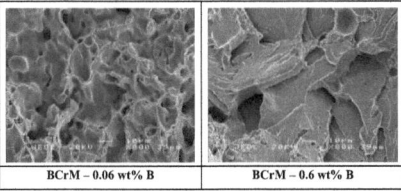

Figure 7: SEM fractography analysis of the BCrM samples sintered for 1h under the protective atmosphere of flowing hydrogen with 0.06 wt% (left) and respectively 0.6 wt% B (right).

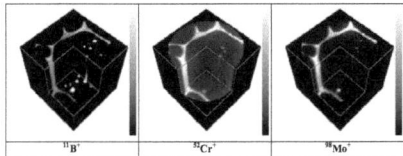

Figure 8: 3D SIMS images of boron, chromium and molybdenum of the sample Fe+3 wt%Cr+1.5 wt%Mo+0.5 wt%C+0.3 wt%B. The homogenous distribution of chromium and molybdenum but also the formation of chromium- and molybdenum-boride is observable. The dimensions of the images: lateral 150μm x 150μm, depth 5 μm. Diameter of the images: 150μm, linear signal intensity: white: 100%, black: 0%.

Figure 9 shows clearly the difference of hardness values for samples sintered without chromium and molybdenum, with 1.5 wt% Cr and 0.2 wt% Mo and respectively with 3 wt% Cr and 0.5 wt% Mo. The values presented for samples without chromium and molybdenum are from formal works and contain a higher carbon content (0.8 wt%).In generally hardness values are increasing at presence of carbon, with higher boron content (higher amount of eutectic phase which is harder than matrix) but it is also limited due to formation of bigger and secondary pores at higher boron content.Again increasing the sintering temperature leads also to slightly enhancement of the hardness values.

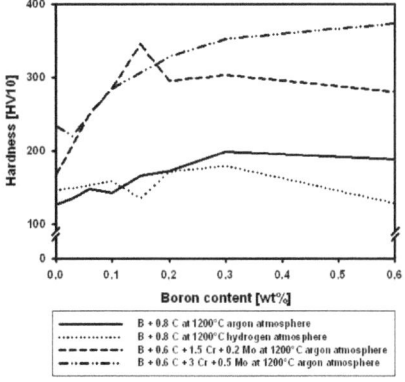

Figure 9: Hardness (HV10) values for samples sintered with and without alloying elements chromium and molybdenum depending on their amount and boron content. Note, that the samples containing Cr and Mo are sintered with lower carbon content (0.6 wt%).

Conclusion

In this work two different prealloyed chromium molybdenum iron powders, CrL (Fe-1.5wt%Cr-0.2wt%Mo) and CrM (Fe-3wt%Cr-0.5wt%Mo) with varying ferroboron amounts were studied. In general, their sintering behaviour and mechanical properties are similar. Boron is a suitable sintering activator in ferrous powder metallurgy up to a certain amount, which depends on some other sintering process parameters. During the sintering process eutectic liquid phase is formed, which is rounding the pores and densifies the material. Larger contents lead to a formation of grain boundary networks and more secondary pores, which can be related to a decrease of the impact energy.
The hardness is not only enhanced by using higher boron amounts but also by sintering with a significant amount of alloying elements. The higher amount of pre alloyed components chromium and molybdenum leads to higher hardness values due to the interaction of these elements with boron during the sintering process.
If sintering is performed with carbon there is

also an enhancement of the hardness values but the disadvantage of the carbon usage is the strong pronounced embrittlement and thus bad ductility of the material. When carbon is added the ferritic microstructure is transformed to a bainitic one and impact energy and hardness are changed. The samples sintered with carbon exhibit from the start on very bad impact toughness behaviour. Probably the post sintering thermal treatments can partly solve this problem.

Sintering under the reductive atmosphere of flowing hydrogen leads to an irregular boron distribution throughout the samples. In the core of the sample specimens the boron distribution depends on its amount: at smaller amounts there are eutectic boron precipitations observable and at higher amounts the grain boundary network is formed. At the surface of the specimens there is an area of approximately 500 µm where boron is strongly depleted. If sintering is performed under the protective atmosphere of argon and with carbon this effect is not observable.

Generally, using higher sintering temperature causes an enhancement of the mechanical properties, due to higher diffusion of all elements and higher formation of sintering necks.

Secondary Ion Mass Spectrometry allows the 2D and 3D elemental distribution analysis as well as of the main components and trace elements in an wide concentration range and offers the results for better understanding of the proceeding processes.

Acknowledgements

The authors want to thank the Austrian Science Fund (FWF) Project P14889 for the financial support for these works.

References

[1] Narasimhan K.S., (2001) Mater Chem Phys (67), 56
[2] German R. M., Madan D. S., (1984) Mod Dev Powder Metall (15), 441
[3] German R. M., (1998) Powder Metallurgy of Iron and Steel, Wiley Chichester, UK
[4] Salak A., (1995) Ferrous powder metallurgy, Cambridge Int Science, Cambridge, UK
[5] Madan D. S., German R. M., James W. B, (1986) Progress in powder metallurgy (42), 19
[6] Madan D. S., (1991) Int J Powder Metall (27/4), 339
[7] Krecar D, Vassileva V, Danninger H, Hutter H, (2004) Anal Bioanal Chem (379), 605
[8] German R.M., Rabin B.H., (1985) Powder Metall (28/1), 7
[9] Selecka M. Salak A., Danninger H., (2003) Mat Proces Techn (143-144), 910
[10] Eisens und Stahlpulver für Sinterformteile, (1998) Höganäs, Sweden
[11] Hutter H., Nowikow K, Gammer K., (2001) Appl Surf Sci (179), 161
[12] Hutter H., (2002) Dynamic Secondary Ion Mass Spectrometry. In: Bubert H., Jenett H. (eds) Surface and thin film analysis. Wiley VCH, Weinheim, Germany, 106
[13] Kazior J., Pieczonka T., Ploszczak J., (2002) Proceedings of the international Conference DF PM 2002, 125
[14] Kazior J., Pieczonka T., Ploszczak J., Nykiel M., (2003) PM 2003, 281

9.4 2D and 3D SIMS Investigations on sintered steels

D. Krecar, J. Zwanziger, V. Vassileva, H. Danninger, H. Hutter

Published in: Applied Surface Science, Elsevier

Volume 252, Number 1, 2005

282 – 285

2D and 3D SIMS investigations on sintered steels

Dragan Krecar, Jürgen Zwanziger, Vassilka Vassileva,
Herbert Danninger, Herbert Hutter *

*Institute of Chemical Technologies and Analytics, University of Technology Vienna,
Getreidemarkt 9/164–AC, A-1060 Vienna, Austria*

Available online 2 March 2005

Abstract

Powder metallurgy (PM) is a well-established method for manufacturing ferrous precision parts. Sintering is one of the important production steps and can be strongly enhanced (activated) by formation of a liquid phase during the sintering process. The liquid phase can be reached by the addition of alloying elements (e.g., copper) or sintering activators (e.g., phosphorus) and is formed by melting of eutectic phase mixtures or by incipient melting. The main investigations presented in this work are done by secondary ion mass spectrometry (SIMS): 2D and 3D elemental distribution. Additionally, impact energy and hardness measurements were performed in order to study the influence of phosphorus on mechanical properties. The concentration of P in different samples was varied between 0 and 1 weight percent (wt.%), the carbon content was consistently 0.5 wt.%. Nominal specimens were sintered at 1120 and 1250 °C in protective atmosphere of flowing nitrogen to determine the influence of sintering temperature.
© 2005 Elsevier B.V. All rights reserved.

PACS: 68.49.Sf; 82.80.Ms; 81.20.Ev

Keywords: 2D–3D SIMS; Sintering; Activator; Phosphorus; PM

1. Introduction

Powder metallurgy (PM) provides a unique opportunity to produce precision components with complex geometry and excellent surface quality. Most applications of sintered steels are in automotive engineering but also non-automotive industry applications are increasing. Sintering can be strongly enhanced by formation of liquid phase during the sintering process. Phosphorus, one of these activating elements, is the focus of this work. Numerous investigations into the iron–phosphorus systems have already been performed [1–7]. Phosphorus addition through Fe_3P powder causes the formation of a transient liquid phase by a eutectic reaction between Fe_3P and Fe during the sintering process at temperatures >1050 °C in protective atmosphere. In this case, the melt is distributed by capillary forces. This liquid phase provides a better distribution of all

* Corresponding author.
 E-mail address: h.hutter@tuwien.ac.at (H. Hutter).

other alloying components and a rounding of the pores and enhances diffusion in α-iron.

This work describes the phosphorus distribution and its influence on impact energy and hardness in PM steels with addition of 0.5 wt.% carbon at two different sintering temperatures. It will be shown that secondary ion mass spectrometry (SIMS) is a very useful technique to investigate 2D and 3D elemental distribution as of trace elements as well as of main components due to its detection limit (sub ppm) and sufficiently lateral resolution (1 μm).

2. Experimental

The investigated sintered samples were produced from water atomized iron powder (ASC 100.29 produced by Höganäs AB, Sweden). All samples (except the reference sample) contain phosphorus acting as sintering activator (added as ferrophosphorus powder Fe_3P, particle size <40 μm) and 0.5 wt.% carbon (by addition of UF 4 natural graphite). The powders were die compacted at 600 MPa and the green compacts were sintered at 1120 and 1250 °C under protective atmosphere of flowing nitrogen. The sample preparation was described in details elsewhere [8,9].

All samples were investigated by means of secondary ion mass spectrometry. The SIMS device used throughout this investigation was an upgraded Cameca IMS 3f [8,10]. Table 1 shows the experimental SIMS parameters for both primary ions beam conditions. In addition to SIMS measurements, impact

Table 1
Experimental secondary ion mass spectrometry setup

	O_2^+ primary ions	Cs^+ primary ions
Beam energy (keV)	5.5	14.5
I_p (μA)	1.5	0.15
Scanned area (μm^2)	300 × 300	300 × 300
Analyzed area (μm ⌀)	150	150
Detected secondary ions	Positive	Negative

The primary ion beam scans rapidly over the sample surface (scanned area) to achieve a homogeneous illumination of the sample (I_p: primary ion current).

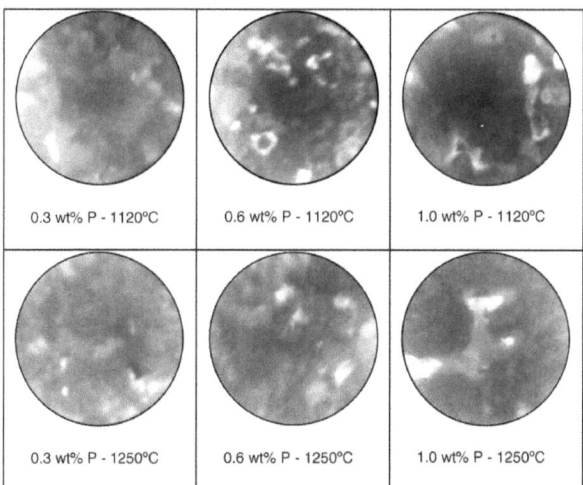

Fig. 1. SIMS 2D images of the phosphorus (detected m/e ratio = 31) with different P amounts sintered at 1120 °C (upper row) and 1250 °C (lower row). Diameter of images: 150 μm; primary ions: Cs^+; beam energy: 14.5 keV; primary ion current: 150 nA; scanned area: 300 μm × 300 μm; linear signal intensity: black = 0%, white = 100%.

toughness (by means of a Charpy impact tester with $W_{Max} = 50$ J on unnotched test specimens, ISO 5754) and hardness (HV 10) measurements were performed.

3. Results and discussion

Both caesium (Cs^+) and oxygen (O_2^+) primary ions were used to identify the elemental distribution of the sintering activator and other trace elements. In this work, the main investigation was done using caesium primary ions for generation and subsequently detection of secondary ions of electronegative elements, e.g., phosphorus, carbon, nitrogen, oxygen, on which this work was focused. Nevertheless, the measurements with oxygen primary ions were also done to detect metallic impurities (sodium, aluminium, silicon, potassium, calcium), which were found at negligible amounts at the grain boundaries or in surrounding pores.

In previous work [8], a homogeneous phosphorus distribution throughout the whole series (0.15–1 wt.%) was demonstrated if no significant carbon amounts were present. Fig. 1 shows the phosphorus distribution at samples containing 0.3, 0.6 and 1 wt.% phosphorus and 0.5 wt.% carbon. Phosphorus is still homogeneously distributed, but with increasing phosphorus concentration the precipitates at grain boundaries and surrounding pores are visible. At higher sintering temperature (1250 °C, Fig. 1 lower row, 1 wt.% P) even a grain boundary network enriched with phosphorus is observable.

The 3D elemental distribution of carbon (Fig. 2) shows the brighter areas, which correspond to pearlite and darker areas corresponding to ferrite. It can be clearly seen that phosphorus is enriched in α-iron. This stabilisation provides higher self-diffusion of iron.

Fig. 3 shows the impact energy and hardness values as a function of P content and sintering temperature. Impact energy values are slightly increased up to 0.15 wt.% P at both sintering temperatures and decrease significantly above 0.15 wt.% P. The decay is more pronounced for specimens sintered at 1250 °C

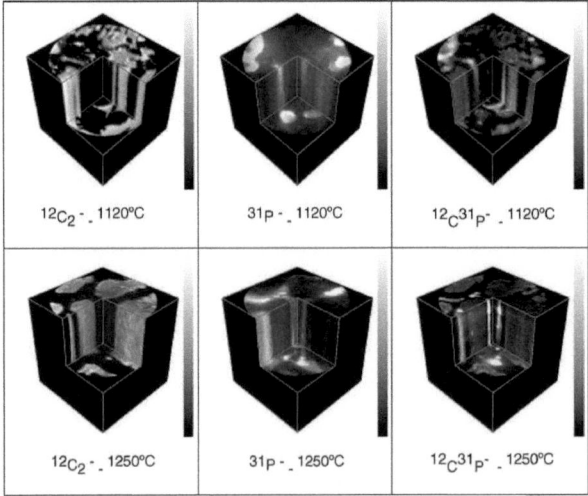

Fig. 2. SIMS 3D images of $^{12}C_2^-$ ($m/e = 24$), $^{31}P^-$ ($m/e = 31$) and $^{12}C^{31}P^-$ ($m/e = 43$) for samples with 1 wt.% P sintered at 1120 °C (upper row) and 1250 °C (lower row). Dimensions: lateral 150 μm × 150 μm, depth 5 μm; primary ions: Cs^+; beam energy: 14.5 keV; primary ion current: 150 Na; scanned area: 300 μm × 300 μm.

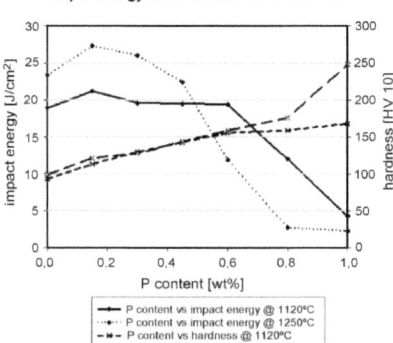

Fig. 3. Influence of phosphorus content and sintering temperature on mechanical properties (impact energy and hardness) of the investigated samples.

because of stronger formation of P precipitates and grain boundary networks at this sintering temperature. As expected, the hardness values increase with increasing P content. If sintering is made at higher temperature (1250 °C), there is also a hardness increase for samples containing 0.8 and above wt.% P.

4. Conclusion

In this work, the distribution of the sintering activator phosphorus and of carbon was investigated by means of 2D and 3D SIMS. It could be shown that beside homogeneously phosphorus distribution with increasing phosphorus content precipitations at grain boundaries and surrounding pores are present, and especially at higher sintering temperature, grain boundary networks of iron and phosphorus are formed. Phosphorus activates the sintering process by an increase in the amount of low-melting phase by eutectic reactions during the sintering process and by stabilization of α-iron. Phosphorus is enriched in α-iron and, if carbon content exceeds 1 wt.%, phosphorus forms a grain boundary network [8].

The hardness increases with increasing phosphorus content due to powerful solid solution hardening effect of phosphorus and hardening effect of carbon but the effect of higher hardness through the higher sintering temperature is only observable for samples with high phosphorus content (0.8 wt.% and above). As expected, the impact energy values are decreasing with increasing phosphorus content because the material becomes more brittle and purely intergranular brittle fracture occurs [9].

It could be shown that secondary ion mass spectrometry is a very useful technique for 2D- and 3D-determination of elemental distribution of sintering activators as well as of trace elements and impurities in PM steels.

Acknowledgement

The authors want to thank the Austrian Science Fund (FWF) – Project P14889 – for the financial support for this work.

References

[1] K.S. Narasimhan, Sintering of powder mixtures and the growth of ferrous powder metallurgy, Mater. Chem. Phys. 67 (2001) 56–65.
[2] H. Miura, Y. Tokunaga, Int. J. Powder Metall. Powder Technol. 21 (4) (1985) 269–280.
[3] R.M. German, D.S. Madan, Mod. Dev. Powder Metall. 15 (1984) 441–454.
[4] A. Molinari, G. Straffelini, V. Fontanari, R. Canteri, Sintering and microstructure of phosphorus steels, Powder Metall. 35/4 (1992) 285–291.
[5] A. Salak, Ferrous Powder Metallurgy, Cambridge International Science Publications, Cambridge, UK, 1995.
[6] A. Molinari, G. Straffelini, R. Canteri, Heat treatment and mechanical behaviour of sintered Fe–C–P Steels, Int. J. Powder Metall. 30/3 (1994) 283–291.
[7] H. Danninger, E. Wolfsgruber, R. Ratzi, in: Proceedings of the Euro PM '97 on Adv. in Structural PM Component Production, Munich, Germany, EPMA, Shrewsbury, 1977, p. 99.
[8] D. Krecar, V. Vassileva, H. Danninger, H. Hutter, Phosphorus as sintering activator in powder metallurgical steels: characterisation of the distribution and its technological impact, Anal. Bioanal. Chem. 379 (2004) 610–618.
[9] V. Vassileva, D. Krecar, C. Tomastik, H. Hutter, H. Danninger, Effect of phosphorus addition on impact fracture behaviour of sintered iron, Fractography (2003).
[10] H. Hutter, Dynamic secondary ion mass spectrometry, in: H. Bubert, H. Jenett (Eds.), Surface and Thin Film Analysis, Wiley–VCH, 2002, pp. 106–121.

9.5 Characterization of wear and surface reaction layer formation on aerospace bearing steel M50 and a nitrogen-alloyed stainless steel

S. Peissl, H. Leitner, R. Ebner, H. Hutter, D. Krecar, R. Rabitsch

Published in: ASTM Special Technical Publication (2007), STP 1465, ASTM International

Bearing Steel Technology: 7th Volume, 214 – 223

Characterization of wear and surface reaction layer formation on aerospace bearing steel M50 and a nitrogen-alloyed stainless steel

Sven Peissl[1], Harald Leitner[2], Reinhold Ebner[1], Herbert Hutter[3], Dragan Krecar[3], Roland Rabitsch[4]

1 Materials Center Leoben, Austria, peissl@unileoben.ac.at
2 Department of Physical Metallurgy and Materials Testing, University of Leoben, Austria
Harald.Leitner@unileoben.ac.at
3 Institute for Chemical Technologies and Analytics, Vienna University of Technologies, Vienna, Austria, herbert.hutter+e164@tuwien.ac.at
4 Boehler Edelstahl GmbH, Kapfenberg, Austria, Roland.Rabitsch@boehler-edelstahl.at

Keywords
Reaction layer formation, Phosphate, TCP, Sliding wear, Mobil Jet II, SIMS

Abstract
The formation of reaction layers on surfaces in mechanical contact is strongly affected by the tribological loading conditions, the materials used, the lubricant and the service temperature. An appropriate balance between reactivity of material and lubricant in tribological systems decreases wear and friction and increases the durability.
Goal of the paper is to compare the reaction layer formation on a standard aerospace bearing steel AMS 6491 (M50) with that on a high strength stainless steel grade AMS 5898 exhibiting a nominal chemical compositions of 0.82C-4.1Cr-1V-4.2Mo (wt.%) and 0.3C-0.4N-15.2Cr-1Mo (wt.%), respectively. As lubricant jet engine oil Mobil Jet II has been used.
Rolling contact fatigue (contact pressure: 6 GPa) and ball-on-disk tests (contact pressure: 1.6 GPa, sliding speed 10 cm/s) at room temperature and at 150°C were employed to study the effect of extreme loading conditions and temperature dependence of the reaction layer formation. The contact areas were inspected by means of optical profilometry, scanning electron microscopy (SEM) and secondary ion mass spectrometry (SIMS) in order to determine type, thickness, homogeneity, and distribution of the reaction layers in the contact zone.
Main result of the study is that the reaction layer formation is significantly less on the stainless steel grade compared with M50. SIMS depth profiles were determined in order to explain the differences in the wear characteristics of the two materials. The reaction layers are mainly built up of P_xO_y molecules, which might be phosphates resulting from the tricresyl phosphate (TCP) lubricant additive. The role of the chemical reaction of the steel and TCP regarding the layer formation will be discussed.

Introduction
The wear of the steels used for standard aircraft engine mainshaft bearings AMS 6491 (M50) and AMS 6278 (M50NiL) is reduced due to the formation of a reaction layer between the surfaces in contact [1]. The layer formation is a result of a chemical reaction between the steel and the oil additive and is strongly affected by the tribological loading conditions, the materials used, the lubricant and the service temperature. An appropriate balance between reactivity of material and lubricant in tribological systems decreases wear and friction and thus increases the durability [2,3].
Mostly in aircraft engine bearings the oil Mobil Jet II, which contains the additive tricresyl phosphate (TCP) illustrated in Fig. 1, is used [4]. TCP underlies a chemisorption to the operating surfaces and thus wear and friction at temperatures up to about 200°C are reduced [5]

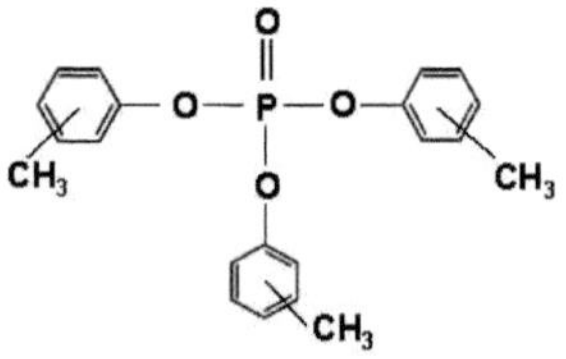

Fig. 1: Structure of tricresyl phosphate.

An investigation conducted by Saba [3] describes the reactivity between TCP and various oxides. In this study oxide powder was placed with the same amount of TCP in a glass ampoule and then heated up to 425°C and held different times between 1h and 24h. Afterwards, the remaining TCP, which did not react with the oxide powder, was analysed. Sabas´s investigations showed that TCP reacts with iron oxide but hardly with chromium oxide. Both standard bearing steels AMS 6491 and AMS 6278 have a chromium content of about 4.1 wt.%. Due to the low chromium content both steel grades have a low corrosion resistance and

the steel surface consists of iron oxides mainly [6]. The iron oxides are beneficial for the reaction with TCP as mentioned above.
However, in recent years the development of new steel grades for bearings show a trend towards increasing chromium content in the matrix in order to guarantee a clearly improved corrosion resistance. The reason for the improvement of the corrosion resistance is a protective passive layer, consisting of chromium oxides, on the steel surface. One example for such a high strength stainless steel grade is the AMS 5898 [7,8,9].
The material development trends and the different reactivity of TCP with oxides reported by Saba were the starting points of this work. However, it is necessary to investigate the effect of changing of steel composition (steel reactivity) in a tribosystem close to reality. The requirements on steels for aircraft engine mainshaft bearings can be divided in a resistance against rolling contact fatigue and against sliding wear (c.p. increasing slip ratios). Therefore, two different testing methods were employed to generate reaction layers on the steel surface. The first one was the ball on disk test, which enables pure sliding conditions and the second one was the ball-rod test, which works under rolling conditions, including varying local micro slip ratios in the contact area.
In order to inspect the contact areas optical profilometry, scanning electron microscopy (SEM) and secondary ion mass spectrometry (SIMS) were employed. The interaction between the steels as mentioned above and the additive TCP will be discussed.

Experimental

The standard aerospace bearing steel AMS 6491 and the high strength stainless steel grade AMS 5898, with the nominal chemical compositions as shown in TABLE 1 were investigated under pure rolling and sliding conditions to compare the reaction layer formation on the surface.
For the experiments all samples were hardened due to austenitizing and ensuing quenching in oil, then follows a deep frozen step, and finally an annealing step with the parameters as given in TABLE 2. These selected parameters assure a constant hardness value of 60 HRC and a sufficient corrosion resistance of AMS 5898.
To study the wear behavior of the steel grades under pure sliding conditions a high temperature ball on disc (BOD) test (CSM Instruments) was conducted.

TABLE 1 – Chemical compositions of the steels investigated [wt.%].

	C	N	Cr	Mo	V	Fe
AMS 6491	0.82	...	4.1	4.2	1	bal.
AMS 5898	0.3	0.4	15.2	1	...	bal.

TABLE 2 – Heat treatment conducted on AMS 6491 and AMS 5898.

	Austenitization temperature [°C]	Deep freeze temperature [°C]	Annealing temperature and time [°C; h]
AMS 6491	1100	-70	585; 3x2h
AMS 5898	1030	-70	185; 2x2h

For this test the heat treated discs, which were grinded and then polished with a 1μm diamond suspension were used. The resulting average roughness of the polished surface was about 5 nm. For each run the sample was completely immersed under a defined amount (7ml, resulting in an oil height over the sample of 5mm) of Mobil Jet II oil. The BOD tests were conducted at room temperature and at 150°C. As sliding contact partner balls with a diameter of 6mm made of AMS 6491 were used. The applied loads were 2 N and 15 N, resulting in a maximum contact stress p_{max} of 834 MPa and 1634 MPa, respectively, calculated by using reference [11].
To compare the reaction layer formation the BOD test with that under rolling contact conditions, rolling contact fatigue (RCF) tests, as described in [10] were performed. In order to carry out the tests at 150°C, the test equipment was slightly modified (Fig. 2). The rod specimens were turned with an oversize on diameter (ø9.93mm), heat treated and finally cylindrically grinded (ø9.525mm). As counterpart for the rolling contact fatigue tests AMS 6491 steel balls were used. According to the applied test load of 113 kg a maximum contact stress of 6418 MPa was reached, calculated by approximate formulae for contact parameters between two elastic bodies [11]. The RCF tests were stopped after 10^6 load contacts. A constant dripping lubrication was supplied (9-11 drops per minute), feeding Mobil Jet II oil.
For inspecting the wear tracks a non-contact optical white light profilometer (NT1000, Veeco Metrology Group) has been used. The measurement setup allows to analyse areas of 598*453nm^2 with a vertical resolution of 3 nanometers for smooth surfaces. The measurement outputs are topography, cross section profile, wear volume, and surface roughness.
To get information about the surface composition of the generated reaction layers, the wear tracks were analysed by scanning electron microscopy (SEM) in back scattering mode and

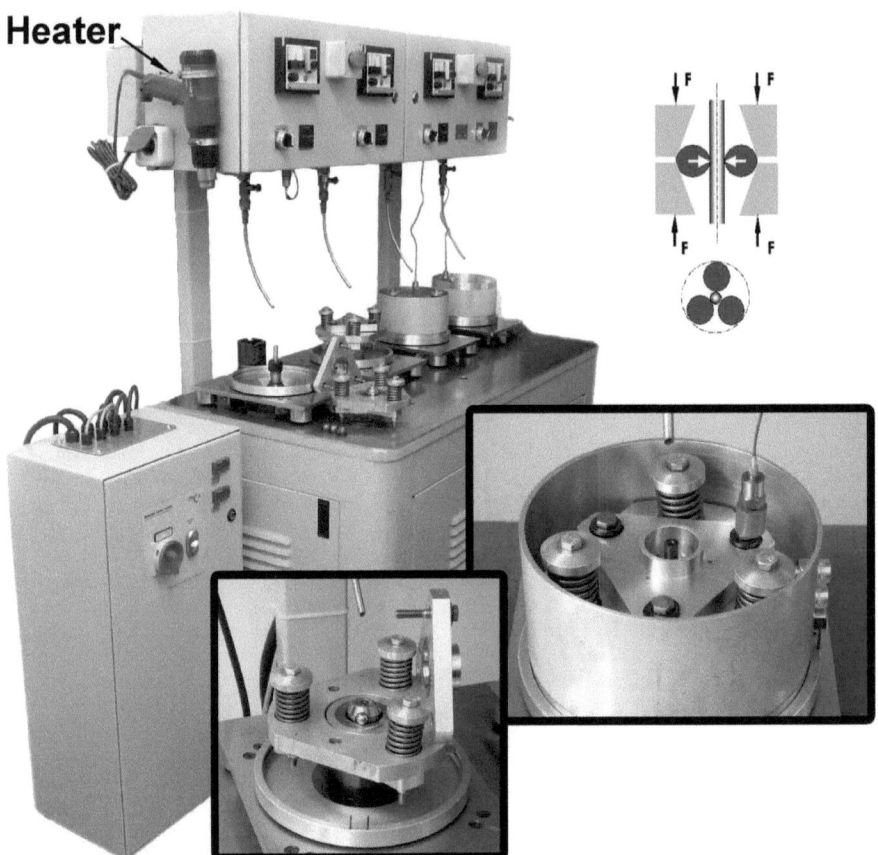

Fig. 2: Modified three ball on rod rolling contact tester for elevated temperature levels.

secondary ion mass spectrometry (SIMS).
The SIMS instrument used throughout this investigation was an upgraded Cameca IMS 3f. The device improvements are mainly in the primary section: an additional primary magnet enables the use of a fine focus Cs^+ ion source as well as a duoplasmatron source (in our case generating O_2^+ primary ions); the original beam deflection was replaced by a digital scan generator.
In stigmatic mode Cameca IMS 3f acts as a light microscope whereby each point on the double channel plate corresponds to one certain point on the sample. Both caesium (Cs^+) and oxygen (O_2^+) primary ions were used to identify the in-depth and lateral elemental distribution of Phosphorus as well as of trace elements. and subsequently detection of electronegative elements, in our case e.g.: phosphorus, carbon, oxygen. In contrast oxygen primary ions are used to generate and detect electropositive elements such as e.g. metals (sodium, aluminium, silicon, potassium, calcium, chromium, iron, molybdenum). Because of possible cluster ion formation during sputtering process nearly all measured masses were also detected with an energy offset to ensure that the measured signal is the desired one. Hereby the sample voltage is decreased and this causes a change in the energy distribution of secondary ions. The mean energy of cluster ions is lower than of atomic ions and this fact is used to reduce the yield of interfering cluster ions by reducing the sample voltage. If sample voltage is lowered more of the atomic ions still have enough energy to pass through an energy slit [12]. This method was also used to improve the non satisfied mass resolution of the used SIMS instrument; e. g.

using 250 V offset the interfering $^{30}Si^1H$ (mass divided by electrical charge m/e = 31) cluster ion disappear completely and the desired ^{31}P (m/e = 31) signal intensity can be measured but with a decreased signal intensity. TABLE 3 shows the experimental SIMS parameters for both primary ions beam conditions.

TABLE 3 – Experimental Secondary Ion Mass Spectrometry setup. The primary ion beam scans rapidly over the sample surface (scanned area) to achieve a homogeneous illumination of the sample.

	O_2^+ primary ions	Cs^+ primary ions
Beam energy	5,5 keV	14,5 keV
Primary Ion current	50 nA	10 nA
Scanned area	250 x 250 µm²	250 x 250 µm²
Analyzed area (diameter)	150 µm	30 µm
Detected secondary ions	Positive	Negative

Result and Discussion

The BOD test gives information about the material behavior under pure sliding conditions. This loading situation is of relevance due to the large amount of slip in aircraft turbine bearings due to great number of revolutions and relatively low load levels.

In order to compare wear and friction behavior at equal pressure and temperature levels (pmax = 1.6 GPa, testing temperature 150 °C), BOD experiments were conducted on AMS 6491 and AMS 5898. After a test run of 1000 m the samples were inspected by means of optical profilometry. A typical image of the surface topography, the corresponding cross sections of wear tracks after the test run and the friction coefficients for the first 20 m sliding distance are shown in Fig. 3a-b. The differences in wear track appearance of the investigated steels are significant. AMS 5898 shows higher wear compared to AMS 6491. The wear track depth at AMS 5898 is in the range of 2.5 µm whereas the wear track depth is only 0.5 µm at AMS 6491.

The friction coefficient µ is in the same range (0.16) after 8 m sliding distance. At shorter distances there are significant fluctuations of µ in the case of AMS 5898. The reason for that is illustrated in Fig. 4. As can be seen in the SEM picture dark areas on the surface are visible. Energy dispersive x-ray spectrometer (EDX) measurements revealed that these areas consist mainly of AMS 5898. Therefore, it can be concluded that material (AMS 5898) was transfered from the disc to the counterpart ball (AMS 6491) and causes an increased friction value at the beginning resulting in an increased wear of the AMS 5898 disc. On this basis the following wear classification steps can be proposed:

1. Material transfer form AMS 5898 disc to ball.
2. Work hardening of transfered particles.
3. Ploughing of AMS 5898 disc due to transfered particles.

Further BOD experiments were conducted to compare the wear behavior at different load and temperature levels. The wear depicted as the standardized wear (volume loss over wear track length) in Fig. 5 was determined after a sliding distance of 1000 m. At the lower load level of 2 N the volume losses of AMS 6491 and AMS 5898 are in the same range. In the case of the higher load level of 15 N the volume loss of AMS 5898 increases significantly.

This might be due to the role of the lubricant film in both cases. If the load is increased to 15 N the lubricant film is not sufficient to separate completely the metallic surfaces in contact and thus the wear mechanism changes. However, at a load of 15N the oil additive TCP should react with the surface to form a protective reaction layer [5]. This layer minimizes wear and friction when asperities of surface come in contact. As the wear of AMS 5898 is 3.5 times higher compared with AMS 6491 at 15 N and 150°C, it seems that the high strength stainless steel grade does not react sufficiently with TCP. Room temperature measurements at the same load showed that wear is constant in the case of AMS 6491 but increased for AMS 5898 compared to the 150°C tests. By considering that the reaction of TCP with the surface needs higher temperatures, a certain pressure and a reactive (steel) surface, this different wear behavior of AMS 6491 and AMS 5898 is explainable. AMS 6491 is reactive enough to form protective reaction layers at 150°C as well as at room temperature. Thus, the wear is in the same the range at room temperature and 150°C. Contrary, the reaction layer formation on AMS 5898 is poor developed due to the high chromium content of the steel. Hence, lowering the test temperature, which results in a reduction of TCP reaction, causes an increased wear.

If the reaction layer formation on the steel is not sufficient, the surfaces of the counterparts will not be separated. Due to the high pressure in the contact zone and the poor reaction layer, adhesive material transfer from AMS 5898 disc to AMS 6491 ball takes place. Consequently, an increased wear behavior of AMS 5898 in the BOD experiments can be observed.

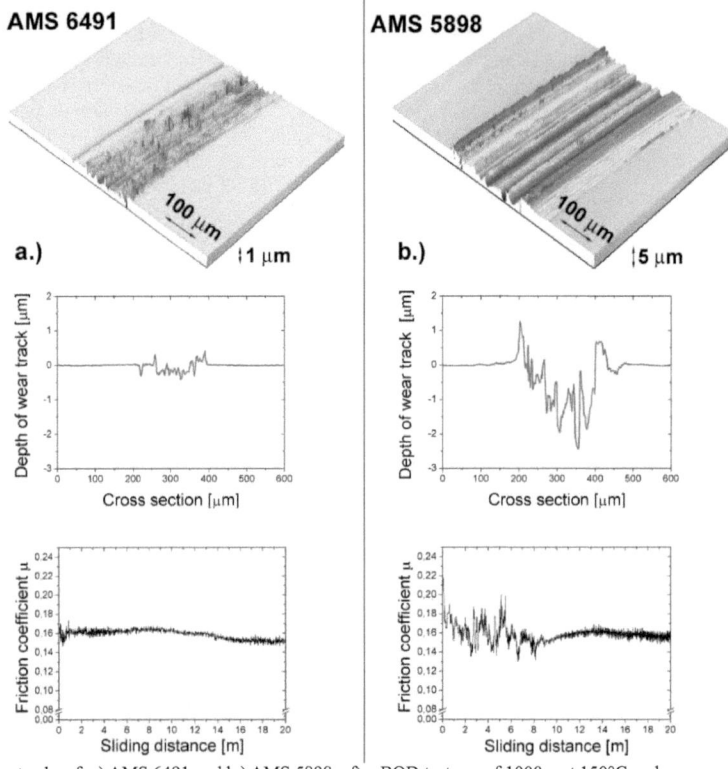

Fig. 3: Wear tracks of a.) AMS 6491 and b.) AMS 5898, after BOD test run of 1000m at 150°C and corresponding cross sections and friction coefficients in the early stages.

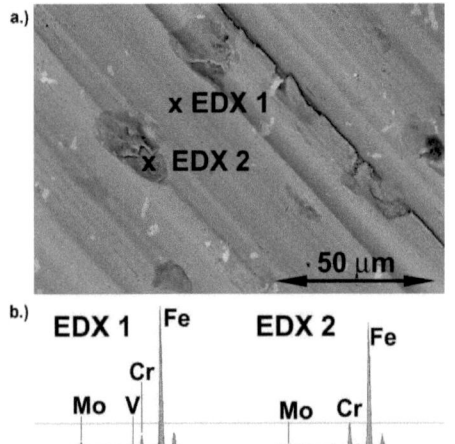

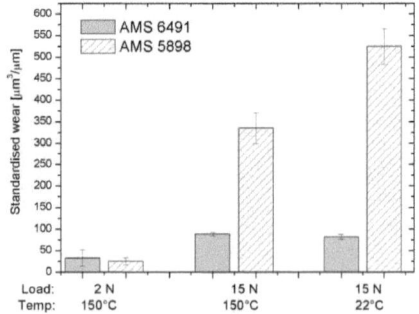

Fig. 5: Standardized volume loss of the discs after BOD test (sliding distance 1000m, speed 10 cm/s) at different loads and temperatures.

Fig. 4: a.) SEM picture of the contact area of AMS 6491 ball after BOD test (8 m sliding distance) against AMS 5898 and b.) corresponding EDX spectra of transfer material (EDX 2) from the disc and ball material (EDX 1).

Further investigations were focused on the characterization of reaction layer formation under rolling conditions with local varying micro slip. Therefore, RCF experiments where performed on both steel grades. Typical SEM pictures of generated tracks on the rods of AMS 6491 and AMS 5898 are shown in Fig. 6a and b. EDX measurements indicates that the dark areas are phosphorus rich regions, being a result of the reaction of TCP with the steel in the contact zone. As can be seen in Fig. 6 there is a clear difference of reaction layer formation of AMS 6491 compared to AMS 5898. Whereas the reaction layer on AMS 6491 is well developed, almost no reaction layer can be observed at AMS 5898.

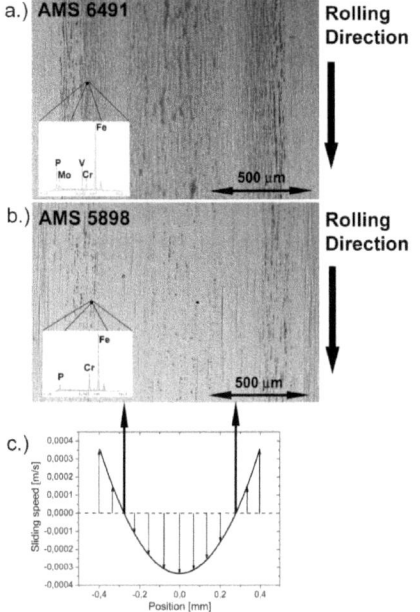

Fig. 6: Typical SEM image of the track after RCF tests (150°C, 106 load alterations) and corresponding EDX spectra. a.) AMS 6491, b.) AMS 5898, c.) Calculated micro slip as a function of the position on the track. The dark areas show the generated reaction layer, which is in conformity with micro slip.

Further, the micro slip behavior over the rod track was calculated by using the geometrical correlations in the contact zone. The assumed rolling motion angle of the ball was 2.5°, starting from the ball center. The course of the calculated micro slip is illustrated in Fig. 6c, which correlates well with the SEM pictures.

By comparing rod track positions, where the calculated slip is zero, the track appears brighter. Consequently, the reaction layer is thinner at these positions. At positions with positive or negative slip the reaction layer is more pronounced developed.

To characterize the reaction layer in more detail, SIMS measurements were employed. The measurements were performed near the center line of the rod tracks, where the reaction layer was sufficiently developed. Measurements on each material were done on two different positions to check the reproducibility. Fig. 7 shows the SIMS phosphorus (m/e =31) profile measured with Cs^+ primary ions and 250V offset voltage. The denoted depth values of the sputter times are approximately depth values. For the calculation of the depth values, the total depth achieved was set in relation with the sputter time, making the assumption, that the rate of the sputtering process is constant. The results indicate significant differences in the depth of the phosphorus containing layers on AMS 6491 and AMS 5898. Within the first 40 min of sputtering (depth ~200nm) AMS 6491 shows a high intensity of P corresponding to the reaction layer. Afterwards a phosphorus enriched region can be observed up to sputtering of 70 min (depth ~350nm).

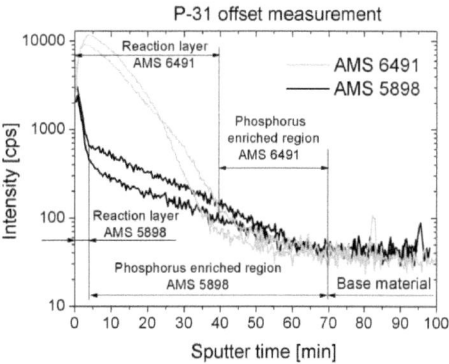

Fig. 7: SIMS phosphorus profiles of AMS 6491 and AMS 5898 (P (m/e = 31); 250 V offset; Cs^+ primary ions).

The measured phosphorus content remains constant for longer sputtering times, which means that the bulk material has been reached. Contrary to this, AMS 5898 shows a well developed reaction layer only up to 4 min sputter time, followed by the phosphorus enriched region, again up to 70 min (depth ~350nm).

To get more information about the distribution of phosphorus on the surface SIMS 2D images has been recorded on AMS 6491 (Fig. 8) and AMS 5898 (Fig. 9). Detected ions and molecule fragments were phosphorus (m/e = 31), and four different kinds of m/e ratios corresponding to PxOy (PO: m/e = 31; PO_2: m/e = 63; PO_3: m/e = 79; and PO_4: m/e = 95). It can be seen that the distribution of phosphorus is congruent with the distribution of P_xO_Y. During the measurement the reaction layer on the steel surface is sputtered away due to the Cs^+ ions and the secondary molecule fragments are detected. It can be expected, that the molecules bonded in the reaction layer, partly are destroyed during sputtering. That is the reason for the measuring of lower m/e ratios like P (m/e = 31) or PO (m/e = 47).

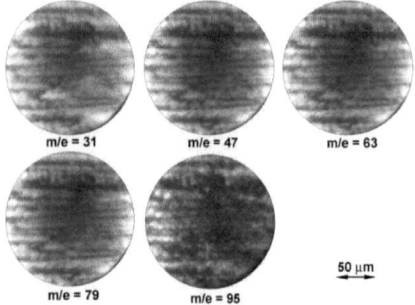

Fig. 8: SIMS 2D images of AMS 6491. Detected m/e ratios were: 31 (P ion distribution), 47 (PO ion distribution), 63 (PO_2 ion distribution) 79 (PO_3 ion distribution), and 95 (PO_4 ion distribution). Linear signal intensity: black = 0%, white = 100%. Primary ions: Cs^+.

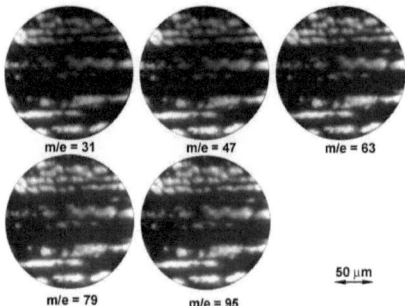

Fig. 9: SIMS 2D images of AMS 5898. Detected m/e ratios were: 31 (P ion distribution), 47 (PO ion distribution), 63 (PO_2 ion distribution) 79 (PO_3 ion distribution), and 95 (PO_4 ion distribution). Linear signal intensity: black = 0%, white = 100%. Primary ions: Cs^+.

However, the main difference between AMS 6491 and AMS 5898 is the difference in the distribution of the measured P and PxOy. So it can be concluded that also the distribution of reaction layer is different. On AMS 6491 the reaction layer is well pronounced, whereas on AMS 5898 not.

Conclusion

For material selection in aerospace bearing applications is it necessary to consider the material reactivity with the additive in the lubricant used. If the reactivity of the steel is too low, increased wear is the consequence. Comparing the reaction layer formation on AMS 6491 and AMS 5898 with TCP the following can be concluded:
AMS 5898 shows a less pronounced reaction layer when compared to the standard aerospace bearing steel AMS 6491.
Due to the less pronounced reaction layer adhesive material transfer under pure sliding conditions from ball to disc at AMS 5898 takes place. This results in ploughing of the surface of the AMS 5898 and increased wear is observed.
The reaction layer formation increases in the case of rolling contact in those areas, where slip is superposed.
The reaction layer on AMS 6491 is evenly distributed near the center line of the running track generated in the RCF test, in the case of AMS 5898 only inhomogeneously and very thin reaction layers are formed.
The reaction layers consist mainly of P_xO_Y molecules, probably PO_4.

---

[1] Godfrey, D: ASLE Trans., 8 (1) (1965) 1
[2] Habig, K.-H. and Feinle, P.: Journal of Tribology, 109, (1987), 569
[3] Saba, C.S. and Forster N.H.: Tribology Letters 12(2), (2002), 135
[4] Winder, C., Balouet, J.-C.: Environmental Research Section A 89, pp. 146-164, (2002).
[5] Stachowiak, G.W. Batchelor, A.W. "Engineering Tribology", ISBN 0-7506-7304-4, pp.84. (2001).
[6] Cornell, R.M., Schwertmann, U., "The Iron Oxides; Structure, Properties, Reactions, Occurrence and Uses", Weinheim VHC, 445-461, (1996).
[7] Berns, H., and Trojahn, W.: Creative Use of bearing steels, ASTM STP 1195, J.J.C. Hoo, Ed., American Society for Testing Materials, West Conshohocken, PA, 1993, pp. 149-155.
[8] Boehmer, H.J., Hirsch, T., and Streit, E.: Bearing steels: Into the 21st century, ASTM STP 1327, J.J.C.Hoo and W.B. Green; Eds., American Society for Testing Materials, West Conshohocken, PA, 1998, 131-151.
[9] Trojahn, W., Streit, E., Chin, H., and Ehlert, D.: Bearing steels: Into the 21st century, ASTM STP 1327, J.J.C.Hoo and W.B. Green; Eds., American Society for Testing Materials, West Conshohocken, PA, 1998, pp. 447-459.
[10] Glover, D.: Rolling Contact Fatigue Testing of Bearing Steel, ASTM STP 771, J.J.C. Hoo, Ed., American Society for testing and Materials, 1982, pp. 107-127.
[11] Hamrock, B.J., Dowson, D., "Ball bearing lubrication, the elastohydrodynamics of elliptical contacts, John Willey & Sons, (1981).
[12] Hutter, H, "Dynamic Secondary Ion Mass Spectrometry", H. Bubert and H. Jenett, Surface and Thin Film Analysis, Wiley VCH, pp. 106-121, (2002).

9.6 SIMS Investigation of Gettering Centres Produced by Phosphorus MeV Ion Implantation

D. Krecar, M. Fuchs, R. Kögler, H. Hutter

Published in: Applied Surface Science, Elsevier

Volume 252, Number 1, 2005

278 - 281

 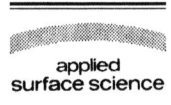

SIMS investigation of gettering centres produced by phosphorus MeV ion implantation

D. Krecar [a], M. Fuchs [a], R. Kögler [b], H. Hutter [a,*]

[a] *Institute of Chemical Technologies and Analytics, TU Vienna, Getreidemarkt 6/164-AC, A-1060 Vienna, Austria*
[b] *Forschungszentrum Rossendorf, P.O. Box 510119, D-01314 Dresden, Germany*

Available online 2 March 2005

Abstract

The ion implantation is a well-known standard procedure in electronic device technology for precise and controlled introduction of dopants into silicon. Damages caused by implantation act as effective gettering zones, collecting unwanted metal impurities. In this work, the consequences of high-energy ion implantation into silicon and of subsequently annealing were analysed by means of secondary ion mass spectrometry (SIMS). The differences in impurities gettering behaviour were studied in dependence of the implantation dose and annealing time at $T = 900\,°C$.
© 2005 Elsevier B.V. All rights reserved.

PACS: 61.72.Tt; 61.72.Yx; 68.49.Sf; 82.80.Ms

Keywords: Ion implantation; Gettering effect and defects; R_P-, $R_P/2$- and trans-R_P-effect; Secondary ion mass spectrometry (SIMS)

1. Introduction

The ion implantation is a well-known standard procedure of electronic device technology for precise and controlled introduction of doping materials into silicon. The process of ion implantation generates damages in material, which still remains after thermal annealing at temperatures of about 900 °C. The residual damage acts as an effective gettering centre for impurities like transition metals in silicon. Such impurities can strongly degrade properties of silicon devices [1].

A gettering layer is formed by ion implantation not only around the mean projected ion range (R_P), but also in the region between surface and R_P. This phenomenon is termed "$R_P/2$-effect" [2]. During a typical MeV ion implantation more than 10^3 silicon atoms are displaced along the trajectory of each implanted ion. Every atom displacement results in one self-interstitial and one vacancy (Frenkel pairs). The separation of vacancies (gettering centres at $R_P/2$) and interstitials (gettering centres at R_P) can be simulated by binary collision models like the well-known computer code transport of ions in matter (TRIM)

* Corresponding author.
 E-mail address: h.hutter@tuwien.ac.at (H. Hutter).

0169-4332/$ – see front matter © 2005 Elsevier B.V. All rights reserved.
doi:10.1016/j.apsusc.2005.02.004

[3]. The third gettering layer is situated in the region deeper than the projected ion range. This phenomenon is called "trans-R_P-effect" [4–6] and is only observed if dopants like P$^+$ or As$^+$ ions are implanted.

The aim of the present work is to investigate the differences in the gettering behaviour of different impurities in silicon after P$^+$ ion implantation and annealing at a temperature of 900 °C. Copper will be used to be the metallic impurity and to mark the defects. Secondary ion mass spectrometry (SIMS) is shown to be an excellent technique for the determination of trace elements, impurities and dopants in silicon with a detection limit in the range of 1 ppm.

2. Experimental

P$^+$ ions were implanted into (1 0 0) p-type Czochralsky (CZ) silicon with the implantation energy of 3.5 MeV. The impact angle was 7°. After the implantation the wafers were than annealed (by furnace annealing) at the temperature of 900 °C for 5 min, respectively, 20 min in an argon ambient. Subsequently, copper was implanted in the backside of the samples. Thereafter, the wafers were heated for 3 min at 700 °C in argon atmosphere to speed up the copper diffusion (diffusivity $D = 5.3 \times 10^{-7}$ cm^2/s at 700 °C [10]) throughout the wafer bulk. An overview of analysed samples is given in Table 1.

The SIMS instrument used throughout this investigation was an upgraded Cameca IMS 3f. The device improvements are mainly in the primary section: an additional primary magnet enables the use of a fine focus Cs$^+$ ion source as well as a duoplasmatron source (for generation of O$_2{}^+$ primary ions). In this work, the samples were investigated by means of Cs$^+$ primary ions (primary energy = 14.5 keV, beam current = 150 nA, scanned area = 350 μm × 350 μm and analysed area = diameter of 60 μm) whereby negative secondary ions were detected [7]. For copper quantification the main copper isotope (^{63}Cu) was monitored. Because of possible cluster ion formation during the sputtering process m/e = 63 ratio was also measured with an energy offset to ensure that measured signal is really copper. Thereby, the sample voltage is decreased and this changes the energy distribution of the secondary ions. The mean energy of cluster ions is lower than of atomic ions and this fact is used to reduce the yield of interfering cluster ions by reducing the sample voltage. If sample voltage is lowered more of the atomic ions still have enough energy to pass through energy slit. Because of non-satisfied mass resolution of the used SIMS instrument this method of reducing of sample voltage is also used for determining of ^{31}P (m/e = 31). Using 250 V offset the interfering ^{30}Si^{1}H (m/e = 31) cluster ions disappear completely, but nevertheless the signal intensity of measured atomic ions decreases. For further sureness ^{63}Cu^{28}Si cluster ion (m/e = 91) was monitored and compared with measured copper signal. The depth profiles of copper and oxygen have been recorded to indicate the gettering sites and phosphorus has been recorded to show projected range.

3. Results

The implantation of the P$^+$ ions (3.5 MeV, 5×10^{15} at/cm^2, sample #1) in CZ silicon and subsequently annealing at 900 °C for 20 min leads to the formation of three gettering layers, which are identified by means of copper depth profiling by SIMS: the $R_P/2$-layer at 1.5 μm, the dominating gettering in the R_P-layer at 2.8 μm, and moreover a copper accumulation in the trans-R_P-layer at approximately 4.3 μm (Fig. 1). The tendency of copper gettering in trans-R_P-layer is only observed from samples implanted with P$^+$ or As$^+$ ions. The copper and oxygen gettering in the R_P-layer as demonstrated at approximately 2.8 μm in Fig. 1 correspond very well with the measured P profile maximum. Phosphorus has been recorded to show the projected range R_P and to compare it with calculates values and the

Table 1
Overview of analysed samples with processing conditions of implantation and annealing step

Sample	Matter	Implantation dose (at/cm^2)	Annealing step (°C/min)
#1	p-Type Si	5×10^{15}	900/20
#2	p-Type Si	5×10^{15}	900/5
#3	p-Type Si	5×10^{14}	900/5

All samples were implanted with the energy of 3.5 MeV and after annealing step copper was implanted in the backside of the samples with implantation energy of 20 keV and an implantation dose of 3×10^{13} at/cm^2.

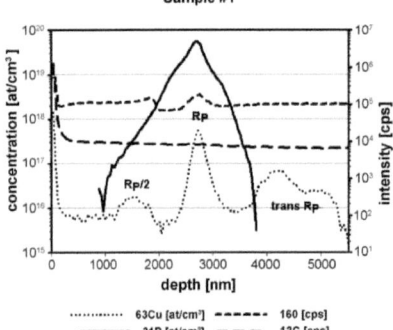
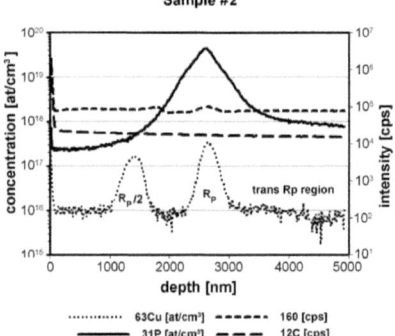

Fig. 1. SIMS depth profile of 3.5 MeV P$^+$ ions (5×10^{15} at/cm^2) implanted ions in CZ silicon (sample #1). The copper distribution shows clearly three effects: copper enrichment at the mean projected ion range (R_P), accumulation at half of the projected ion range ($R_P/2$) and also copper enrichment beyond the projected ion range (trans-R_P).

Fig. 2. Copper, phosphorus, oxygen and carbon profiles of the sample #2 measured by SIMS in CZ silicon implanted with P$^+$ ions (3.5 MeV, 5×10^{15} at/cm^2). The sample was annealed at 900 °C for 5 min.

maxima of P depth profile agree very well with the projected range calculated with TRIM. The defect free zone is between 1.8 and 2.3 μm.

Shortening the annealing time from 20 to 5 min (sample #2) a slightly different copper distribution is observed. In comparison with the 20 min annealed sample (Fig. 1) the getter efficiency of the $R_P/2$-layer is stronger pronounced in this sample and no copper gettering in the region beyond the R_P range is observable (Fig. 2).

If P$^+$ ions are implanted with lower dose of 5×10^{14} at/cm^2 but with same implantation energy of 3.5 MeV and also annealed for 5 min as described above a completely other copper gettering behaviour is obtained as shown in Fig. 3.

4. Discussion and conclusions

Using P$^+$ for implantation causes always, as expected, copper and oxygen gettering at the dislocations, which are present in the R_P-layer after annealing. For gettering at cavities in the $R_P/2$-layer the prolongation of the annealing time from 5 to 20 min leads to a stronger gettering effect for O, whereas Cu gettering is reduced. This different behaviour reflects a different trapping mechanism. At 700 °C copper atoms are almost homogenous distributed ($D = 5.3 \times 10^{-7}$ cm^2/s at 700 °C [10]) throughout the wafer bulk and subsequently copper diffusion caused by cooling is trap-limited [11]. Implanting P$^+$ a higher number of gettering centres is

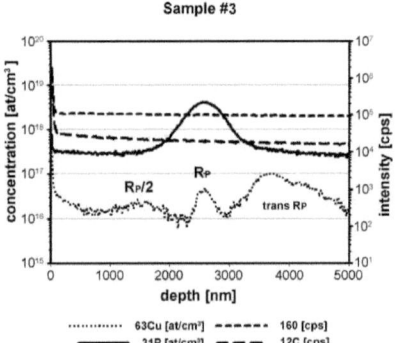

Fig. 3. Copper, oxygen, carbon and phosphorus profiles measured by SIMS in CZ silicon implanted with 3.5 MeV P$^+$ ions with ion fluence of 5×10^{14} at/cm^2 and annealed at 900 °C for 5 min (sample #3).

build at R_P and their number increases during annealing time. The copper gettering in the trans-R_P gettering layer (consists of still unknown defects, probably of interstitial clusters) is only observed after precursory longer annealing step. It may be concluded that the formation of these defects is related with the phosphorus diffusion mechanism [5,8]. Figs. 1–3 show also a slightly copper accumulation at the wafer surface because of the natural sink behaviour of surface for impurities. This indicated copper accumulation at the wafer surface can be overlapped with an artefact at the beginning of SIMS measurements. The gettering in region beyond the mean projection range (trans-R_P-effect) is more pronounced for longer annealing time and for lower P^+ implantation dose. Higher implantation doses create at the projection range more damages, which act as strong sink for impurities [6,9].

Longer annealing time forces better oxygen diffusivity ($D = 1.8 \times 10^{-12}$ cm^2/s [10]), and therefore the oxygen getter efficiency is increased in the region of the R_P/2-layer whereas the copper gettering efficiency is decreased. Carbon, as the second intrinsic impurity, which was monitored through out this work, shows no significant gettering behaviour probably because of the very strong bond between carbon and silicon (SiC).

Secondary ion mass spectrometry has shown its outstanding faculty to detect metals and trace elements with the detection limit in the range of at least 1 ppm.

The quantification of copper and phosphorus with standards fortifies the superiority of SIMS versus other analytical methods for this kind of problems.

References

[1] T.E. Seidel, Gettering in silicon, in: M. Wittmer, J. Stimmell, M. Strathman (Eds.), Material Issues in Silicon Integrated Circuit Processing, Materials Research Society, Pittsburgh, 1986, pp. 3–12.
[2] M. Tamura, T. Ando, K. Ohya, Nucl. Instrum. Methods Res. B 59–60 (1991) 572.
[3] J.B. Biersack, L.G. Haggmark, Nucl. Instrum. Methods Phys. Res. B 174 (1980) 257.
[4] Y.M. Gueorgiev, R. Koegler, A. Peeva, D. Panknin, A. Muecklich, R.A. Yankov, W. Skorupa, Appl. Phys. Lett. 75 (1999) 3467.
[5] Y.M. Gueorgiev, R. Koegler, A. Peeva, A. Muecklich, D. Panknin, R.A. Yankov, W. Skorupa, J. Appl. Phys. 88 (2000) 5465.
[6] R. Koegler, A. Peeva, A. Lebedev, M. Posselt, W. Skorupa, G. Oezelt, H. Hutter, M. Behar, J. Appl. Phys. 94 (7) (2003) 3834.
[7] H. Hutter, Dynamic secondary ion mass spectrometry, in: H. Bubert, H. Jenett (Eds.), Surface and Thin Film Analysis, Wiley/VCH, 2002, pp. 106–121.
[8] Y.M. Gueorgiev, R. Koegler, A. Peeva, D. Panknin, A. Muecklich, R.A. Yankov, W. Skorupa, Appl. Phys. Lett. 75 (1999) 3467.
[9] Y.M. Gueorgiev, R. Koegler, A. Peeva, A. Muecklich, D. Panknin, R.A. Yankov, W. Skorupa, J. Appl. Phys. 88 (10) (2000) 5645.
[10] D.J. Fischer, Diffusion in Silicon, Scitec Publications, 1998.
[11] N.E.B. Cowern, Appl. Phys. Lett. 64 (20) (1994) 2646–2648.

9.7 Investigation of Gettering Effects in CZ-type Silicon with SIMS

D. Krecar, M. Fuchs, R. Kögler, H. Hutter

Published in: Analytical and Bioanalytical Chemistry, Springer

Volume 381, Number 4, 2005

1526 – 1531

D. Krecar · M. Fuchs · R. Koegler · H. Hutter

Investigation of gettering effects in CZ-type silicon with SIMS

Received: 25 October 2004 / Revised: 17 January 2005 / Accepted: 21 January 2005 / Published online: 7 April 2005
© Springer-Verlag 2005

Abstract Ion implantation is a well-known standard procedure in electronic device technology for precise and controlled introduction of dopants into silicon. However, damage caused by implantation acts as effective gettering zones, collecting unwanted metal impurities. This effect can be applied for "proximity gettering" reducing the concentration of impurities in the active device region. In this study the consequences of high-energy ion implantation into silicon and of subsequent annealing were analysed by means of secondary ion mass spectrometry (SIMS). Depth profiles were recorded of such impurities as copper, oxygen and carbon to obtain information about their gettering behaviour. The differences in impurities gettering behaviour were studied as a function of the implanted ions, P and Si, of the implantation dose and annealing time at $T=900°C$. Besides impurities gettering at the mean projected range (Rp) of implanted ions, Rp-effect, defects at around half of the projected ion range, Rp/2-effect, and even in some cases beyond Rp, trans-Rp-effect, have also been found to be effective in gettering of material impurities.

Keywords Ion implantation · Gettering effect and defects · Rp-effect · Rp/2-effect · Trans-Rp-effect · Secondary ion mass spectrometry (SIMS)

D. Krecar · M. Fuchs · H. Hutter (✉)
Institute of Chemical Technologies and Analytics,
TU Vienna, Getreidemarkt 6/164–AC,
1060 Vienna, Austria
E-mail: h.hutter@tuwien.ac.at
Tel.: +43-1-5880115120
Fax: +43-1-5880115199

R. Koegler
Forschungszentrum Rossendorf,
P.O. Box 510119, 01314 Dresden, Germany

Introduction

Ion implantation is a well-known standard procedure of electronic device technology for precise and controlled introduction of dopants into silicon. The process of ion implantation generates damage in the crystalline semiconductor material, which still remains after thermal annealing at temperatures of 700–1,000°C. This residual damage acts as an effective gettering centre for impurities like transition metals in silicon. Such impurities can strongly degrade properties of silicon devices [1]. The problem worsens as the size of devices is scaled down. Impurity gettering can be applied to collect unwanted metal impurities and reduce their concentration in the device areas. This is referred to as "proximity gettering" [2]. High-energy ion implantation in the MeV range can be applied to getter the metal impurities in a buried layer slightly deeper than the device region. The main advantage of this structure is the location of gettering layer close to the active device area. In this way an effective impurity trapping can also be achieved for low thermal budgets required in advanced device technologies, which limits the diffusion length of dopants and impurities. A promising example of proximity gettering is He^+ ion implantation. It causes [3] formation of cavities in silicon. Cavities trap metal impurities on their inner walls by chemisorption [4].

A gettering layer is formed by ion implantation not only around the mean projected ion range (Rp), but also in the region between surface and Rp: this phenomenon is termed the Rp/2-effect [5]. During a typical MeV ion implantation more than 10^3 silicon atoms are displaced along the trajectory of each implanted ion. Each atom displacement results in one self-interstitial and one vacancy (Frenkel pairs). The radiation-induced vacancies and interstitials are assumed to recombine locally during annealing. This process leads to a spatial separation of the generated vacancies and self-interstitials resulting on average in a vacancy-rich region at Rp/2 and an excess of interstitials in the Rp region [6–8]. With the exception of the implanted atoms (+1 atoms), only the point defects remain which are in (local) excess. The separation of vacancies and interstitials can be simulated by binary collision models like the well-known computer code transport of ions in matter (TRIM) [9]. The gettering sites for impurities around Rp/2 are ascribed to the excess vacancies and the gettering sites at Rp to excess interstitials. The excess interstitials around Rp form interstitial loops and dislocations during annealing, which can be easily observed by cross section transmission electron microscopy (XTEM) [10–12]. The excess vacancies cluster to empty cavities; however, their detection needs special TEM specimen preparation as shown recently [13]. Moreover, there is a third gettering layer in the region deeper than the projected ion range. This phenomenon is called the trans-Rp-effect and is only observed if dopants like P^+ or As^+ ions are implanted [11, 14, 15]. The trans-Rp defects form during annealing by dopant and point defect diffusion. The gettering centres in the trans-Rp region are not yet detected by TEM.

The aim of the present study is to investigate the differences in the gettering behaviour of different impurities in silicon after P^+ and Si^+ ion implantation and annealing at a temperature of 900°C. Oxygen and carbon impurities are introduced in silicon during crystal growth and are always present in Czochralsky silicon (CZ-Si). Metal impurities are introduced in silicon mostly during processing of wafers in device production. Such impurities are very detrimental to essential device characteristics such as carrier lifetime and reverse current. In these experiments copper was chosen as metal impurity. On one hand it is applied as the interconnector material in advanced device technology instead of Al because of its higher conductivity [16]; on the other hand, copper is dangerous because of its extremely high diffusion rate in silicon. Furthermore, Cu is also unstable in perfect crystalline silicon at room temperature (RT) and is accumulated at crystal defects (or at the surface) [17]. This characteristic makes it promising in detection of very small defects by profiling using secondary ion mass spectrometry (SIMS). SIMS is shown to be an excellent technique for the determination of trace elements, impurities and dopants in silicon with a detection limit in the range of 1 ppm.

Experimental

Sample preparation

Si^+ (isotope ^{28}Si) and P^+ (^{31}P) were implanted into (100) p-type CZ-Si with an implantation energy of 3.5 MeV. The impact angle was 7°, and the implantation dose was 5×10^{15} atoms cm^{-2}, except from one P^+-implanted sample (#5, Table 1) with 5×10^{14} atoms cm^{-2}. After the implantation the wafers were cut into small samples and annealed (by furnace annealing) at 900°C for 5 or 20 min in an argon atmosphere. This annealing step is necessary to remove the defects created by the implantation process. Subsequently copper was implanted in the backside of the samples with implantation energy of 20 keV and an implantation dose of 3×10^{13} atoms cm^{-2} (sample thickness 0.5 mm; diffusivity $D(Cu) = 4.3\times10^{-5}$ cm^2 s^{-1} at 700°C [18]). Thereafter, the wafers were heated for 3 min to 700°C in an argon atmosphere to speed up the copper diffusion throughout the wafer bulk. An overview of analysed samples is given in Table 1.

SIMS measurements

The SIMS device used throughout this investigation was an upgraded Cameca IMS 3f. The device improvements are mainly in the primary section: an additional primary magnet enables the use of a fine-focus Cs^+ ion source as well as a duoplasmatron source (in our case generating O_2^+ primary ions); the original beam deflection was replaced by a digital scan generator. Within the scope of this work, Cs^+ ions accelerated to energies of 14.5 keV, resulting in a primary current of 150 nA, and focussed to a spot diameter of 50 µm have been used for the generation of subsequently detected negative secondary ions. The primary ion beam was projected onto an area of 350×350 µm^2 with an analysed area of 150 µm in diameter selected by the use of an aperture diaphragm [19].

Quantification

From the measured depth profiles of the respective standards (copper implantation standard 1.4-MeV/6×10^{13}

Table 1 Summary of sample preparation with processing conditions of implantation, annealing step and subsequently contamination with copper

Sample no.	Matter	Implantation dose (atoms cm^{-2})		Energy (MeV)	Process steps	
		P^+	Si^+		Annealing	Dose(atoms cm^{-2}) $^{63}Cu^+$
1.	p-Type Si	–	5×10^{15}	3.5	900°C/20 min	3×10^{13}
2.	p-Type Si	–	5×10^{15}	3.5	900°C/5 min	3×10^{13}
3.	p-Type Si	5×10^{15}	–	3.5	900°C/20 min	3×10^{13}
4.	p-Type Si	5×10^{15}	–	3.5	900°C/5 min	3×10^{13}
5.	p-Type Si	5×10^{14}	–	3.5	900°C/5 min	3×10^{13}

atoms cm^{-2} and phosphorus homogenous standard 4.5×10^{18} atoms cm^{-3}), relative sensitivity factors (RSF) for copper (^{63}Cu) and phosphorus (^{31}P) regarding silicon (^{30}Si) as reference mass were calculated. With these sensitivity factors we were able to quantify respective impurity concentrations in our samples. In addition to ^{63}Cu and ^{30}Si, the elements ^{12}C and ^{16}O, which can also be gettered at the defect centres as well as the molecule ion $^{63}Cu^{28}Si$, to get a second signal for the copper trend, were measured. The depth scales have been determined by using a SLOAN DEKTAK II A profilometer with an accuracy of about ± 3 rel.%. Sputter-induced roughening effects may also decrease the depth accuracy to some extent, though no significantly higher roughness was found in the crater bottom.

Results

The result of a TRIM calculation for the implantation of 3.5-MeV Si^+ ions into silicon at an impact angle of 7° shows Rp at 2.6 μm. Figure 1 shows the SIMS depth profiles of three impurities in sample#1 (3.5 MeV implanted Si^+ ions, 5×10^{15} atoms cm^{-2}). The dominating copper gettering in p-type CZ-Si is at the region of the mean projected range of the implanted silicon ions whose maximum is at about 2.9 μm, but there is also a significant amount of copper in the region at half of the projected ion range at 1.5 μm, Rp/2-layer. A defect-free zone (free of impurity accumulation) is positioned between 2.0 and 2.3 μm. The copper enrichment at the surface is partly an artefact of the SIMS measurements; however, the surface is also a natural sink for the impurities. Shortening the annealing time from 20 to 5 min (sample#2) causes a slightly different copper distribution. Compared with the 20-min annealed sample (Fig. 1), the getter efficiency of the Rp/2-layer is more pronounced in this sample. The oxygen trapping is stronger in Fig. 1(sample#1) than in Fig. 2 (sample#2). The oxygen concentration in the defect-free zone between 1.8 and 2.3 μm is lower in Fig. 1 than the average oxygen concentration of the substrate, measured by means of FTIR, of about 3×10^{17} atoms cm^{-3}. The missing O is accumulated in the two gettering layers at Rp/2 and Rp. The Cu and O profile maximum of the Rp layer agrees, whereas for the Rp/2-layer the gettering peaks of O are slightly shifted by about 200–250 nm toward deeper positions. The C depth profile seems not to be affected by the Si implant.

After the implantation of the P^+ ions (3.5 MeV, 5×10^{15} atoms cm^{-2}, sample#3) in CZ-Si, copper gettering in three layers is clearly observable: the Rp/2-layer at 1.5 μm, the dominating gettering in the Rp-layer at 2.8 μm and moreover a copper accumulation in the trans-Rp-layer at approximately 4.3 μm (Fig. 3). The tendency of copper gettering in the trans-Rp layer is only observed from samples implanted with P^+ ions not for Si^+-implanted ones. The copper and oxygen gettering in the Rp-layer as demonstrated at approximately

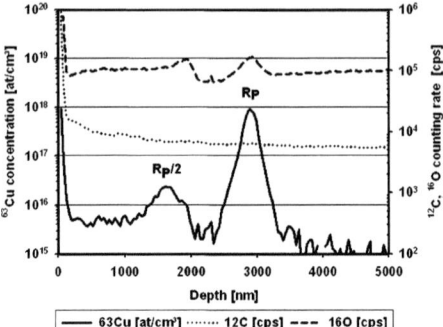

Fig. 1 SIMS depth profile of sample#1. 3.5-MeV Si^+ ions (5×10^{15} atoms cm^{-2}) were implanted, then annealed at 900°C for 20 min and subsequently contaminated with copper by implantation into the rear side and redistribution throughout the sample by an additional thermal treatment at 700°C for 3 min. Beside quantified Cu distribution (*left axis*), this figure also shows the non-quantified in-depth distribution of carbon and oxygen (*right axis*)

2.8 μm in Fig. 3 and at 2.7 μm in Fig. 4 agrees very well with the measured P profile maximum. The P depth profile fits very well with the TRIM calculation. Its depth distribution is slightly broadened because of P diffusion during annealing. As in the case of Si^+ implantation for the P^+-implanted samples, the Cu gettering in the Rp/2-layer is much more pronounced for the shorter annealing time (5 min, Fig. 4, sample#4), whereas no copper gettering in the region beyond the Rp range is observable (Fig. 4). The defect-free zone is similar to the silicon-implanted sample#2 between 1.8 and 2.3 μm. In the Rp/2 gettering layer the profile

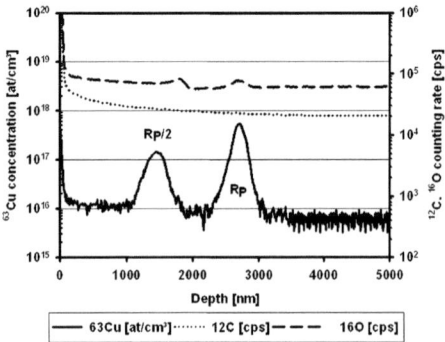

Fig. 2 SIMS depth profiles of sample#2. The parameters are the same as for sample#1 but the annealing time was shorter (only 5 min). In addition to the quantified Cu distribution (*left axis*), this figure shows the non-quantified in-depth distribution of carbon and oxygen (*right axis*)

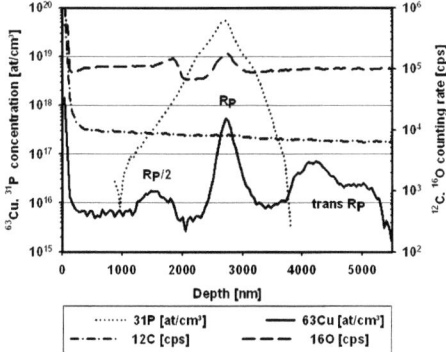

Fig. 3 SIMS depth profile of 3.5-MeV P$^+$ ions (5×10^{15} atoms cm^{-2}) implanted in CZ-Si (sample#3). The sample was annealed at 900°C for 20 min. The copper distribution shows three clear effects: copper enrichment at the mean projected ion range (Rp), accumulation at half of the projected ion range ($Rp/2$) and also copper enrichment beyond the projected ion range ($trans$-Rp). Phosphorus and copper distributions are quantified and the data belong to the *left axis*; the carbon and oxygen graphs are non-quantified and belong to the *right axis*

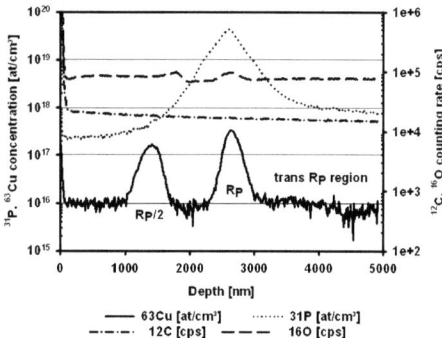

Fig. 4 Copper, phosphorus, oxygen and carbon profiles of sample#4 measured by SIMS in CZ-Si implanted with P$^+$ ions (3.5 MeV, 5×10^{15} atoms cm^{-2}). The sample was annealed at 900°C for 5 min

maxima of Cu and O do not coincide. The O profile is shifted in Figs. 3 and 4 to deeper positions, as in the case of the Si implant (Figs. 1, 2). The gettering efficiency of the Rp/2 region increases for O and decreases for Cu with longer annealing times. The C distribution is not affected.

Figure 5 shows a different copper gettering behaviour if P$^+$ ions are implanted with a lower dose of 5×10^{14} atoms cm^{-2} but with the same implantation energy of 3.5 MeV and also annealed for 5 min as described above for sample#4. A small Cu enrichment is visible in the Rp/2 layer at 1.5 μm and relatively weak Cu gettering appears at the projected ion range at 2.6 μm, but the dominating copper gettering is observed at the region beyond (deeper than 3.6 μm) the projected ion range in the trans-Rp gettering range. The oxygen in-depth distribution shows no gettering neither in the Rp/2-layer nor in the region of the Rp-layer.

Discussion

Table 2 gives an overview of the obtained results throughout this study. The use of Si$^+$ for implantation causes always, as expected, copper and oxygen gettering at the dislocations, which are present in the Rp-layer after annealing. For gettering at cavities in the Rp/2-layer the prolongation of the annealing time from 5 min to 20 min leads to a stronger gettering effect for O, whereas Cu gettering is reduced. This different behaviour reflects a different trapping mechanism. Oxygen is mobilized by the annealing step and its trapping is diffusion-limited. The Cu is introduced after annealing and redistributed by a unique thermal treatment at 700°C. The Cu distribution reflects the distribution of gettering centres. Cavities at Rp/2 undergo Ostwald ripening during thermal treatment, they increase in size and decrease in their number density. This might be the main reason for the reduced gettering efficiency after longer annealing. Additionally it should be taken into account that the O gettering at cavities suppresses the gettering of Cu [20]. The small shift of the O profile in comparison to the Cu profile indicates an internal difference of the structure of Rp/2 defects. The origin of this difference is probably the size distribution of cavities versus depth. Further TEM investigations will verify this

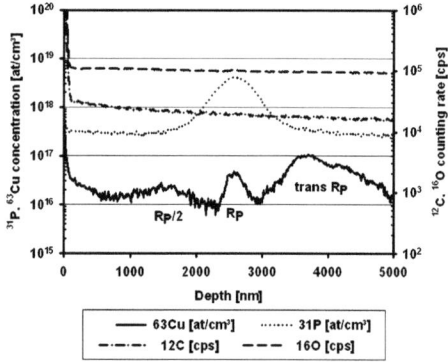

Fig. 5 Copper, oxygen, carbon and phosphorus profiles measured by SIMS in CZ-Si implanted with 3.5-MeV P$^+$ ions with ion fluence of 5×10^{14} atoms cm^{-2} and annealed at 900°C for 5 min. Phosphorus and copper distributions are quantified (*left axis*); the carbon and oxygen graphs are non-quantified (*right axis*)

Table 2 Summary of results for oxygen and copper gettering (maximum concentrations in the gettering layer) as a function of primary implanted ions, annealing time and dose of implanted ions

+ gettering, – no gettering

Sample no.	Implanted ions	O gettering in the ... layer		Cu gettering in the ... layer		
		Rp/2	Rp	Rp/2 (atoms cm^{-3})	Rp (atoms cm^{-3})	Trans-Rp (atoms cm^{-3})
1.	Si$^+$	+	+	2.36×10^{16}	9.05×10^{17}	–
2.	Si$^+$	+	+	1.47×10^{17}	5.33×10^{17}	–
3.	P$^+$	+	+	1.80×10^{16}	5.47×10^{17}	7.22×10^{16}
4.	P$^+$	+	+	1.62×10^{17}	3.43×10^{17}	–
5.	P$^+$	–	–	2.56×10^{16}	4.87×10^{16}	1.04×10^{17}

assumption. No gettering is observed in the region beyond the projected range for Si$^+$ implantation.

The experiments performed with P$^+$ ion implantation show the same behaviour like the Si$^+$ implant regarding gettering at the Rp/2-layer and Rp-layer; however, in contrast to the Si$^+$ implant there is also gettering in the regions beyond the projection ion range, trans-Rp. Unlike Rp/2-layer (cavities) and Rp-layer (dislocations), the trans-Rp gettering layer consist of still unknown defects (probably of interstitial clusters), whose formation proceeds during the annealing treatment. It may be concluded that the formation of these defects is related to the phosphorus diffusion mechanism [14, 15].

The, in general, stronger copper gettering in the Rp-layer compared to the Rp/2-layer is obviously independent of both primary implanted ions. Two effects could explain this: (a) the higher number of gettering centres at Rp [21] and their increase during annealing for longer time (20 min) at 900°C, whereas the concentration of gettering centres at Rp/2 is lower [21] and decreases during anneal and (b) the used CZ-Si has a rather high oxygen amount (>10^{18} atoms cm^{-3}). At the rather high temperature of 900°C, small oxygen precipitates possibly grow during annealing, reducing the open volume for other impurities by emission of interstitials [22]. The intrinsic oxygen impurity is always gettered both at the projected ion range (Rp-layer) and at about half of the projected range (Rp/2-layer). The oxygen gettering is a competitive process to metal gettering at the cavities. For the longer annealing time the oxygen getter efficiency is increased in the region of the Rp/2-layer, whereas the copper gettering efficiency is decreased. This different behaviour and the observed shift in the depth position of the O and Cu profile can be explained by the assumption that the corresponding gettering centres are not exactly the same ones. Cu gettering seems to be related to the concentration of cavities, and O gettering is related to the open volume (concentration of vacancies). The concentration of cavities decreases during annealing and the concentration of vacancies remains essentially constant [22].

The second intrinsic impurity, which was monitored throughout this work, carbon, shows no significant gettering behaviour probably because of the very strong bond between carbon and silicon (SiC).

Conclusion

SIMS has demonstrated its outstanding ability to detect metals and trace elements with a detection limit in the range of at least 1 ppm. The quantification of copper and phosphorus done by correlating their respective measured signals with the values of standards with well known copper and phosphorus concentrations fortifies the superiority of SIMS over other analytical methods for this kind of problem. The experimental data obtained by SIMS can help with the development of further models to explain the formation and effectiveness of formed gettering layers.

References

1. Seidel TE (1986) In: Wittmer M, Stimmell J, Strathman M (eds) Gettering in silicon. Material issues in silicon integrated circuit processing. Materials Research Society, Pittsburgh, pp 3–12
2. Wong H, Cheung NW, Chu PK, Liu J, Mayer JW (1998) Appl Phys Lett 52:1023
3. Fichtner P, Behar M, Kaschny J, Peeva A, Koegler R, Skorupa W (1999) Appl Phys Lett 77(7):972
4. Myers SM, Follstaedt DM (1996) J Appl Phys 79(3):1337
5. Tamura M, Ando T, Ohya K (1991) Nucl Instrum Meth Res B 59–60:572
6. Brown RA, Kononchuk O, Rozgonyi GA, Koveshnikov S, Knights AP, Simpson PJ, Gonzalez F (1998) J Appl Phys 84:2459
7. Venezia VC, Eaglesham DJ, Haynes TE, Agarwal A, Jacobson DC, Gossmann H-J, Baumann FH (1998) Appl Phys Lett 73:2980
8. Holland OW, Xie L, Nielsen B, Zhou DS (1996) J Electron Mater 25(1):99
9. Biersack JB, Haggmark LG (1980) Nucl Instrum Meth Phys Res B 174:257
10. Peeva A, Koegler R, Brauer G, Werner P, Skorupa W (2000) Mater Sci Semicond Process 3:297–301
11. Koegler R, Peeva A, Lebedev A, Posselt M, Skorupa W, Oezelt G, Hutter H, Behar M (2003) J Appl Phys 94:3834
12. Koegler R, Peeva A, Anwand W, Brauer G, Werner P, Gösele U (1999) Appl Phys Lett 75/9:1279–1281
13. Peeva A, Koegler R, Skorupa W (2003) Nucl Instrum Meth Res B 206:71–75
14. Gueorguiev YM, Koegler R, Peeva A, Panknin D, Muecklich A, Yankov RA, Skorupa W (1999) Appl Phys Lett 75:3467
15. Gueorguiev YM, Koegler R, Peeva A, Muecklich A, Panknin D, Yankov RA, Skorupa W (2000) J Appl Phys 88:5465

16. Muraka SP, Gutmann RJ, Kaloyeros AE, Lanford WA (1993) Thin Solid Films 236:257
17. Flink Ch, Feick H, McHugo SA, Seifert W, Hieslmair H, Heiser Th, Istratov A, Weber ER (2000) Phys Rev Lett 85:4900
18. Heiser T, Mesli A (1993) Appl Phys A 57:325–328
19. Hutter H (2002) In: Bubert H, Jenett H (eds) Dynamic secondary ion mass spectrometry. Surface and thin film analysis. Wiley/VCH, New York, pp 106–121
20. Koegler R, Peeva A, Anwand W, Werner P, Danilin AB, Skorupa W (1999) Solid State Phenomena 69–70:235
21. Koveshnikov S, Kononchuk O (1998) Appl Phys Lett 73:2340
22. Koegler R, Peeva A, Werner P, Skorupa W, Gösele U (2001) Nucl Instrum Meth B 175–177:340

9.8 Study of defect engineering in the initial stage of SIMOX processing

R. Kögler, A. Mücklich, L. Vines, D. Krecar, A. Kuznetsov, W. Skorupa

Published in: Nuclear Instruments and Methods in Physics Research,
                Section B: Beam Interactions with Materials and Atoms,
                Elsevier

                Volume 257, 2007

                161 – 164

Study of defect engineering in the initial stage of SIMOX processing

R. Kögler [a,*], A. Mücklich [a], L. Vines [b], D. Krecar [c], A. Kuznetsov [b], W. Skorupa [a]

[a] *Forschungszentrum Dresden-Rossendorf, Bautzner Landstrasse 128, 01328 Dresden, Germany*
[b] *University of Oslo, Department of Physics, POB 1048 Blindern, 0316 Oslo, Norway*
[c] *TU Wien, Inst. für Chem. Tech. und Analytik, Getreidemarkt, 1060 Wien, Austria*

Available online 8 January 2007

Abstract

Defect engineering for SiO$_2$ precipitation was investigated during ion beam synthesis in the first stage of SIMOX processing. Vacancy defects were created in Si: (i) by a buried nanocavity layer pre-fabricated by He implantation and annealing and (ii) by excess vacancy generation during oxide synthesis induced by an additional simultaneous high-energy Si implantation.
A narrow nanocavity layer was found to be an excellent nucleation site that effectively assists SiO$_2$ formation. Such cavity layer must be adjusted to the excess vacancy profile of the O implant. The excess vacancy generation by simultaneous dual implantation avoids defect formation in Si. However, it is inappropriate to form a narrow oxide layer due to the too broad distribution of excess vacancies.
© 2007 Elsevier B.V. All rights reserved.

PACS: 61.72.Tt; 61.72.Yx; 61.72.Ji

Keywords: Ion implantation; SIMOX process; Vacancy defects; Defect engineering; Si

1. Introduction and motivation

Ion beam synthesis of a buried oxide layer in silicon is commonly known as the SIMOX process (separation-by-implantation-of-oxygen), a well established technique to fabricate silicon-on-insulator (SOI) structures. The process consists of two steps: the high-fluence O implantation at temperatures ⩾ 500 °C and the long-term high-temperature annealing at temperatures ⩾ 1300 °C that results in a box-shaped O depth profile (BOX). Defect engineering facilitates nucleation and growth of SiO$_2$ precipitates by introduction of additional nucleation sites localizing the implant and by creation of defects preferring the desired reaction. Recent investigations demonstrated the effect of defect engineering for the case of SiC synthesis [1,2]. The key effect was the introduction of vacancy defects. The presence of vacancies enables the recombination of released interstitials and strain relaxation. For Si oxidation into SiO$_2$ the volume increases dramatically, by about 120%, as compared with the rather small figure of 2.4% for SiC [1]. Therefore, defect engineering by introduction of vacancy defects is expected to be an excellent technique for improvement of the SIMOX process [3]. However, the beneficial effect of vacancies only persists as the growing SiO$_2$ precipitates are fully surrounded by crystalline Si. Volume expansion results in simple surface swelling when a continuous oxide layer develops. For that reason the influence of defect engineering should be restricted to the initial stage of SIMOX processing whilst SiO$_2$ precipitates are disconnected.

In this study two different methods of defect engineering are applied. First, a cavity layer (CL) was pre-fabricated by He implantation and annealing, and secondly, excess vacancies were generated "in-situ" during O implantation by simultaneous dual implantation (DI) [4] of O$^+$ and Si$^+$ ions. The effect of defect engineering was investigated at the initial stage of SIMOX processing as well as after the final BOX anneal.

---
* Corresponding author. Address: Forschungszentrum Dresden-Rossendorf, FWIM, PF 510119, 01314 Dresden, Germany. Tel.: +49 351 260 3613; fax: +49 351 260 3411.
  E-mail address: koegler@fz-rossendorf.de (R. Kögler).

0168-583X/$ - see front matter © 2007 Elsevier B.V. All rights reserved.
doi:10.1016/j.nimb.2006.12.253

2. Experimental

CZ-Si(1 0 0) wafers, p-type with a resistivity of $\geqslant 1\,\Omega$ cm, were pre-implanted with 45 keV He^+ to 4×10^{16} cm^{-2} and subsequently annealed at 800 °C for 30 min to form a buried CL. DI was performed with 200 keV O^+ to fluences of 1×10^{17} and 3×10^{17} cm^{-2} and with 1.1 MeV Si^+ to fluences between 3.5×10^{16} and 6.5×10^{16} cm^{-2} at 550 °C under 22.5° with respect to the surface normal. The mean projected ion ranges (R_P) for DI of O and Si were 0.425 and 1.42 μm, respectively. A pre-anneal at 850 °C for 5 h or at 1100 °C for 2 h to initiate SiO_2 precipitation was followed by oxidation (wet) at 1150 °C for 7 min and by deposition of a 1 μm thick oxide capping layer. Samples were annealed at 1250 °C for 5 h and in a final step at 1350 °C for 5 h (BOX anneal). Reference samples were prepared without DI and CL.

O depth profiles were measured by secondary ion mass spectrometry (SIMS). The defect structure was analyzed by cross sectional transmission electron microscopy (XTEM) and the local resistivity of the buried oxide layer, the main functional parameter of a SIMOX structure, was monitored by cross sectional scanning spreading resistance microscopy (SSRM) [5,6].

3. Results and discussion

The parameters chosen for O implantation are inadequate to manufacture a continuous oxide layer as they are off the optimum dose-energy match which is about 6×10^{17} O^+/cm^2 for 200 keV ion energy [7]. O^+ ion implantation under 22.5° also results in a wider O distribution than for normal incidence. Isolated SiO_2 precipitates in Si (the initial stage of SIMOX) are observed also after the BOX anneal, and hence the effect of defect engineering on nucleation and growth of SiO_2 precipitates can be investigated up to such high temperatures as applied for BOX anneal without complete coalescence of the precipitates. The way to introduce vacancy defects into Si before and during O implantation is demonstrated in Fig. 1. First, nanocavities with maximum sizes of 25 nm were created by He^+ ion pre-implantation and annealing. He diffuses out during annealing and during O implantation [8]. The buried 125 nm wide CL was adjusted in order to match with the maximum O profile. Secondly, excess vacancies were generated simultaneously with O implantation by DI. Vacancy production by simultaneous Si^+ ion implantation continues when pre-fabricated nanocavities disappeared during O implantation. The Si implant generates excess vacancies up to the depth of 1 μm whereas excess interstitials are generated beyond 1 μm. A layer of dislocation loops formed by the Si implant may act as an interstitial trap well separated from the region of oxide synthesis. There is another useful effect of simultaneous Si implantation that is ion beam induced epitaxial crystallization (IBIEC) [9]. Fig. 2 shows how IBIEC avoids defect formation in crystalline Si around SiO_2 precipitates whereas this

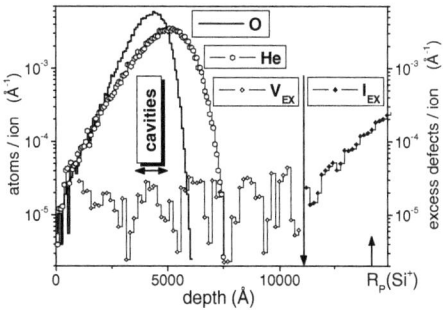

Fig. 1. Calculated depth profiles of a 200 keV O implant (22.5°) and a 45 keV He implant. The position is indicated of the He induced 125 nm wide cavity layer. The excess vacancy profile (V_{EX}) and the excess interstitial profile (I_{EX}) generated by a 1.1 MeV Si implant (22.5°) were calculated as described in [11]. The arrow marks the transition from the excess vacancy to excess interstitial region.

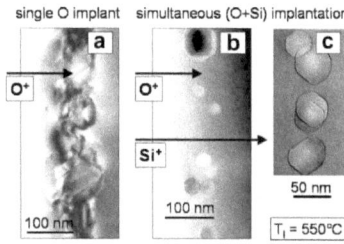

Fig. 2. XTEM bright field images of samples with single O implant (a) and with simultaneous O and Si implantation (b) to fluences of 1×10^{17} cm^{-2} and 5×10^{16} cm^{-2}, respectively, after annealing at 1250 °C. SiO_2 precipitates are shown in higher magnification in a under-focus image (c). All samples contained He induced nanocavities in Si.

region is full of twins and stacking faults for the reference sample (Fig. 2(a)). However, XTEM images (not shown) reveal the appearance of small SiO_2 precipitates beyond the R_P band. The effect of a nanocavity layer alone is demonstrated in Fig. 3 by O profiles measured after annealing at 1250 °C. The pre-fabricated CL is visible in the XTEM image inserted in Fig. 3(b). Calculated excess vacancy and excess interstitial profiles generated by the O implant itself are shown in Fig. 3(a). These vacancies also assist the SiO_2 precipitation. Comparing the excess vacancy concentrations by Si implantation (Fig. 1) and by O implantation (Fig. 3(a)) and considering the corresponding ion fluences one finds that the Si implant contributes not more than about 15% to the total generated excess vacancy concentration. Excess vacancies originated by the O implant are located above the maximum of O profile at a depth $x\leqslant 0.4$ μm. Therefore, the O profile in Fig. 3(b) is not as narrow as allowed by the nanocavity layer. The cavity layer is nevertheless an excellent seed for nucleation and growth of SiO_2 precipitates. In Fig. 4 results are presented for

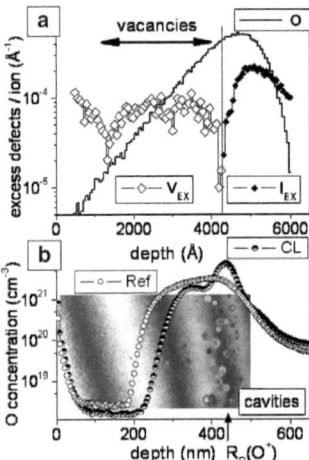

Fig. 3. O depth profile for 200 keV O implantation under 22.5° and the corresponding excess vacancy (V_{EX}) and excess interstitial profile (I_{EX}) calculated as described in [11] (a). Measured O profiles after implantation with ion fluence of 1×10^{17} cm^{-2} and annealing at 1250 °C for defect engineering by a cavity layer (CL) and for the reference sample, i.e. without defect engineering (Ref). The TEM image shows He induced nanocavities in Si after annealing at 850 °C (b).

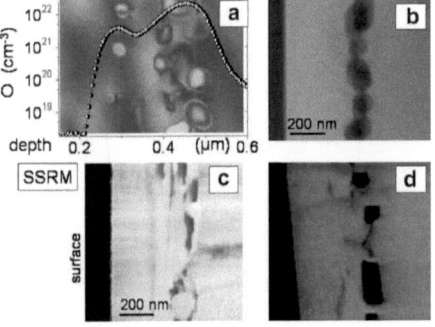

Fig. 4. TEM images (a, b) of nanocavity containing samples implanted with 3×10^{17} O$^+$/cm^2, 200 keV after annealing at temperatures of 1250 °C (a, c) and 1350 °C (b, d). Resistivity scans 1×1 μm^2 (SSRM) taken from sample cross sections (b, d). Dark areas represent high resistivity. The surface oxide is left hand. The O profile after annealing at 1250 °C is given (a).

Table 1
Mean width of the buried oxide layer, mean size of oxide precipitates and oxide area of the cross section determined from SSRM 1×1 μm^2 scans for samples with simultaneous dual implantation (3×10^{17} O$^+$/cm^2 and 6.5×10^{16} Si$^+$/cm^2) and a cavity layer (DI + CL) and for samples only with cavity layer (CL)

Anneal	1250 °C		1350 °C	
Sample	DI + CL	CL	DI + CL	CL
Mean width (nm)	208	215	182	161
Mean size (nm)	1680	1560	10,565	7955
Oxide area (nm^2)	37,700	41,100	70,800	52,100

reported in the literature [10] and is ascribed to SiO$_2$ precipitation at the positions of O profile maximum and damage maximum. In view of Fig. 3 the excess vacancy generation, and not the total damage, is the origin of the second oxygen band. The adjustment of the cavity layer to the O profile maximum proves to be unfavorable to achieve a narrow oxide layer. The local resistivity of sample cross sections is shown by 1×1 μm^2 scans after annealing at 1250 °C (Fig. 4(c)) and 1350 °C (Fig. 4(d)). High resistivity areas are ascribed the SiO$_2$ precipitates. The observed SiO$_2$ precipitates appear smaller then they actually are and such ones with sizes smaller than the diamond tip ($\leqslant 10$ nm) cannot be resolved by SSRM. The details of SSRM analysis are summarized in Table 1. After annealing at 1250 °C a significant amount of O is still in small SiO$_2$ precipitates with sizes below the detection limit. This is demonstrated by the increase of resistivity and oxide area (Table 1) during BOX annealing which results in a significant growth and coalescence of SiO$_2$ precipitates.

4. Conclusions

The introduction of a nanocavity layer is shown to assist SiO$_2$ precipitation in the first stage of SIMOX processing. However, the cavity layer must be in good match with the excess vacancy profile generated by the O$^+$ ion implantation to achieve a narrow oxide layer. Moreover, the cavity layer has to be tailored not to be wider than the proposed BOX profile.

Excess vacancy generation by an additional simultaneous implantation does not result in faster SiO$_2$ precipitate growth and coalescence. The mean size of precipitates and the width of the precipitate layer is not significantly affected after annealing at 1250 °C, otherwise the mean size and the amount of precipitates increases during BOX anneal at 1350 °C. That indicates the excess vacancy assisted formation of stable small SiO$_2$ precipitates during simultaneous dual implantation. These precipitates may form in the whole region of the excess vacancy distribution also outside the proposed BOX layer. They need high temperatures for their resolution to make possible further precipitate growth in the oxide region. The advantage of simultaneous dual implantation is mainly to avoid implantation defects in crystalline Si around SiO$_2$ precipitates. However, such

defect engineering by nanocavities after annealing at 1250 °C (left) and 1350 °C (right). After 1250 °C the O profile (Fig. 4(a)) typically shows two O peaks: one at the position of the O profile maximum (R_P peak), and another one at shallower position, $x \leqslant 0.4$ μm. The shallower O peak always appears notwithstanding the R_P peak is enforced by the CL. Such dual oxide band structure was also

defects usually are removed during SIMOX processing by the high temperature BOX anneal at temperatures $\geqslant 1300\ °C$.

References

[1] R. Kögler, F. Eichhorn, J.R. Kaschny, A. Mücklich, H. Reuther, V. Heera, W. Skorupa, C. Serre, A. Perez-Rodriguez, Appl. Phys. A 76 (2003) 827.
[2] R. Kögler, F. Eichhorn, A. Mücklich, W. Skorupa, C. Serre, A. Perez-Rodriguez, Vacuum 78 (2005) 177.
[3] R. Kögler, A. Mücklich, H. Reuther, D. Krecar, H. Hutter, W. Skorupa, Solid State Phenom. 108–109 (2005) 321.
[4] J.R. Kaschny, R. Kögler, H. Tyrrof, W. Bürger, F. Eichhorn, A. Mücklich, C. Serre, W. Skorupa, Nucl. Instr. and Meth. A 551 (2005) 200.
[5] P. De Wolf, J. Snauwaert, T. Clarysse, W. Vandervorst, L. Hellemans, Appl. Phys. Lett. 66 (1995) 1530.
[6] P. Eyben, T. Janssens, W. Vandervorst, Mater. Sci. Eng. B 124–125 (2005) 45.
[7] M. Chen, X. Wang, J. Chen, Y. Dong, W. Yi, X. Liu, Xi Wang, J. Vac. Sci. Technol. B 21 (2003) 2001.
[8] W. Fukarek, J.R. Kaschny, J. Appl. Phys. 86 (1999) 4160.
[9] F. Priolo, E. Rimini, Mater. Sci. Rep. 5 (7–8) (1990) 319.
[10] A. Ogura, H. Ono, Appl. Surf. Sci. 159–160 (2000) 104.
[11] R. Kögler, A. Peeva, J.R. Kaschny, W. Skorupa, H. Hutter, Nucl. Instr. and Meth. B 186 (2002) 298.

9.9 Low energy RBS and SIMS analysis of the SiGe quantum well

D. Krecar, M. Rosner, M. Draxler, P. Bauer, H. Hutter

Published in: Applied Surface Science, Elsevier

Volume 252, Number 1, 2005

123 – 126

Low energy RBS and SIMS analysis of the SiGe quantum well

D. Krecar [a], M. Rosner [a], M. Draxler [b], P. Bauer [b], H. Hutter [a],*

[a] *Institute of Chemical Technologies and Analytics, TU Vienna, Getreidemarkt 6/164 AC, A-1060 Vienna, Austria*
[b] *Institut fuer Experimentalphysik, Johannes Kepler Universitaet Linz, Altenbergerstrasse 69, A-4040 Linz, Austria*

Available online 19 April 2005

Abstract

The Ge concentration in a MBE grown SiGe and the depth of the quantum well has been quantitatively analysed by means of low energy Rutherford backscattering (RBS) and secondary ion mass spectrometry (SIMS). The concentrations of Si and Ge were supposed to be constant, except for the quantum well, where the nominal germanium concentration was of 5%. Quantitative information was deduced out of raw data by comparison to SIMNRA simulated spectra. With the knowledge of the response function of the SIMS instrument (germanium delta (δ) layer) and using the model of forward convolution (point to point convolution) it is possible to determine the germanium concentration and the thickness of the analysed quantum well out of raw SIMS data.
© 2005 Published by Elsevier B.V.

PACS: 68.49.Sf; 82.80.Ms; 82.80.Yc; 73.21.Fg; 68.18.−g; 68.47.Pe

Keywords: Secondary ion mass spectrometry; SIMS; Low energy Rutherford backscattering; RBS; Quantum well; Ge-δ-layer

1. Introduction

Improving the performance of semiconducting materials, the construction of new materials like heterogeneous structures of silicon and germanium and its products is one of the main research areas of the semiconductor industry. The Si/SiGe heterostructures (single or multiple quantum wells) are one of the newer developed materials, which are used for the production of electronic devices (e.g. heterojunction bipolar transistors, HBTs) [1], but also for the production of quantum cascade lasers (near- to far-

* Corresponding author.
E-mail address: h.hutter@tuwien.ac.at (H. Hutter).

infrared wavelength ranges and semiconductor diode laser) [2]. The Si/SiGe heterostructures (interfaces) are preferably manufactured by means of molecular beam epitaxy (MBE).

With its good depth resolution and the advantage of standard-free quantification RBS is a widely used quantitative technique to analyse the thickness of a film in the nanometre range. SIMS is widely used as a depth profiling technique with some drawbacks caused by sputter processes. The combination of raw data with mathematical fitting procedures e.g. convolution with the known response function (measured germanium delta monolayer) allows the determination of layer parameters such as thickness, position, and element concentration of the quantum well.

0169-4332/$ – see front matter © 2005 Published by Elsevier B.V.
doi:10.1016/j.apsusc.2005.01.109

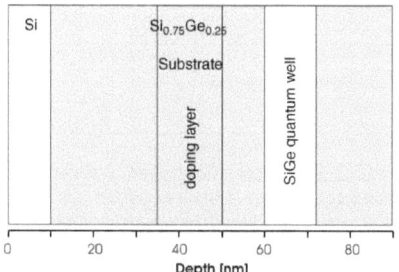

Fig. 1. Layer structure of analysed samples. Si_xGe_{1-x} (quantum well) heterostructure is situated at 60 nm (nominal). In the doping layer Sb is added in low concentration.

2. Experimental

The MBE production of SiGe samples was described in detail elsewhere [3]. The samples consist of different layers as shown in Fig. 1. The nominal Ge concentration in the quantum well was 5%. In the other layer Si and Ge concentration is supposed to be constant: $Si_{0.75}Ge_{0.25}$. The cap layer consists only of silicon.

The substrate of the Ge-δ-layer was a Czochralski-grown Si p-type wafer. One atomic layer of Ge was deposited onto the substrate with MBE and additionally 50 nm Si was deposited onto the Ge layer.

The used SIMS instrument was an upgraded Cameca IMS 3f. The primary ion beam was projected onto an area of 350 μm × 350 μm in square with an analysed area of 60 μm in diameter selected by the use of an aperture diaphragm [4]. The recorded masses were $^{16}O^+$, $^{30}Si^+$ and $^{70}Ge^+$ and accompanying with all measurements the response function of the SIMS device (Ge-δ-layer) was detected.

The RBS experiments were carried out at the −15° beamline of the 700 keV van de Graaff accelerator at the institute of experimental physics at the Johannes Kepler Universität Linz. The experimental setup is described in detail in [5].

3. Results and discussion

Rutherford backscattering (RBS) is one of the mostly used techniques for non-destructive, reference-free quantitative analysis of composition, thickness, and depth profiles of thin films or interfaces [6,7]. Typically H^+ or He^+ MeV ions are accelerated to the target and the energy of the backscattered ions is analysed. The use of computer simulation programs like SIMNRA [8] is the most effective way to calculate RBS spectra. Knowing the optimal measuring para-

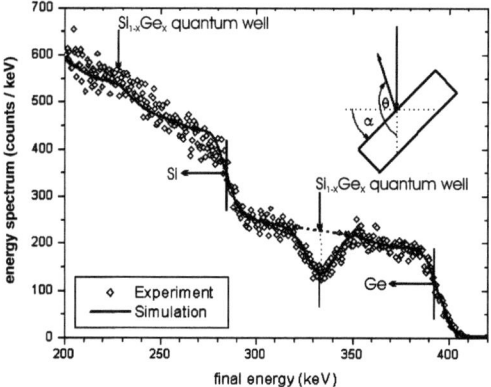

Fig. 2. Measured and SIMNRA simulated RBS spectra for sample with 5% Ge for 500 keV He ions and an angle of incidence $\alpha = 45°$.

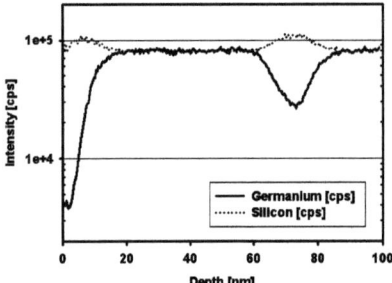

Fig. 3. SIMS depth profile of germanium and silicon of the investigated sample. The used primary ions were O_2^+ with an energy of 5.5 keV. The angle of incidence was 42.4°. The primary ion beam was scanned over an area of 350 μm × 350 μm with a primary ion current of 30 nA, secondary ions were collected from the centre (diameter 60 μm) of scanned area.

meters the samples were analysed by means of low energy RBS. After the energy calibration of the multichannel analyser the RBS spectra of the samples were recorded at 500 keV. Fig. 2 shows a simulated and a measured RBS spectrum for the sample with nominal 5% Ge in the quantum well. Because of the thicker Si cap layer than expected (nominal value 10 nm, measured value 13 nm), the position of the Ge high-energy edge shifts to lower values (Fig. 2).

Table 1
Comparison of low energy RBS and SIMS results

Point of investigation	Nominal value	RBS	SIMS
Ge concentration in the QW (%)	5	4	4
Thickness of the QW (nm)	12	12.4	12.1
Position of the QW (nm)	At 60	At 63.0	At 64.9

Fig. 3 shows a SIMS in depth profile of Si and Ge of the sample with nominal Ge concentration of 5%. The decrease of the Ge concentration in the quantum well, as seen in the RBS spectrum (Fig. 2), is clearly visible. In Fig. 4, the Ge-δ-layer and appropriate SIMS depth profile are shown (lower part of Fig. 4: Ge-δ-layer nominal and Ge-δ-layer SIMS measurement). The SIMS depth profile of the Ge-δ-layer is the response function of the SIMS instrument. By means of this response function the further calculation of the Ge concentration, position, and thickness of the quantum well is possible. For this procedure the profile of the Ge-δ-layer was normalized and convolved point to point with the model profile. The resulting convolution was manually fitted to the measured Ge slope trying to vary parameters until the best match of both graphs was reached. The quantitative RBS measurements of Ge concentration were used as standards for further SIMS measurements and quantification. The result obtained for a sample with nominal 5% Ge in the quantum well is displayed in Fig. 4 (upper part of Fig. 4: SIMS depth profile, model and convolution) and Table 1.

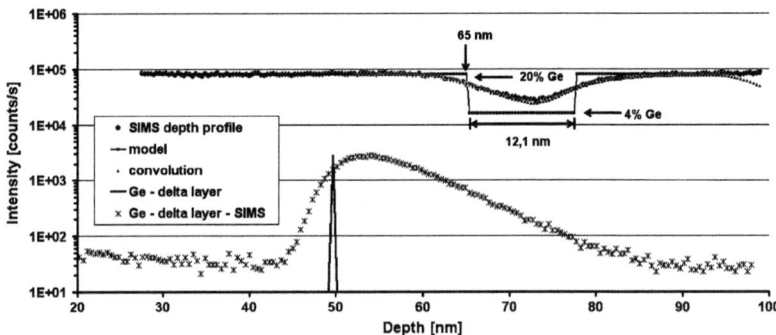

Fig. 4. Demonstration of the convolution procedure. A model is convolved point to point with the SIMS response function, Ge-δ-layer, and modified until the best agreement with the measured profile is achieved. The channel width is determined with 12.1 nm (RBS: 12.4 nm).

4. Conclusion

In this work the Ge concentration in MBE grown SiGe heterostructures was investigated by means of SIMS and low energy RBS. It could be demonstrated that both techniques are definitely able for quantitative analysis of Ge in the SiGe quantum well. The results of SIMS agree very well with the results obtained with low energy RBS and there is also a well agreement with nominal values (Table 1). It could be shown that the drawback of atomic mixing at SIMS, which strongly limits the depth resolution, can be compensated at the data evaluation by means of mathematical procedures, point to point convolution, as used in this work.

References

[1] J.M. Fernández, L. Hart, X.M. Zhang, M.H. Xie, J. Zhang, B.A. Joyce, J. Cryst. Growth 164 (1996) 241.
[2] M. Maier, D. Serries, T. Geppert, K. Köhler, H. Güllich, N. Herres, Appl. Surf. Sci. 203–204 (2003) 486.
[3] F. Schäffler, D. Többen, H.-J. Herzog, G. Abstreiter, B. Holländer, Semicond. Sci. Technol. 7 (1992) 260.
[4] H. Hutter, Dynamic secondary ion mass spectrometry, in: H. Bubert, H. Jenett (Eds.), Surface and Thin Film Analysis, Wiley/VCH, 2002, pp. 106–121.
[5] F. Kastner, PhD Thesis, Johannes Kepler Universität, Linz.
[6] W.K. Chu, J. Mayer, M.-A. Nicolet, Backscattering Spectrometry, Academic Press, New York, 1978.
[7] J.R. Tesmer, M. Nastasi, Handbook of Modern Ion Beam Materials Analysis, Materials Research Society, Pittsburgh, 1995
[8] M. Mayer, Nucl. Instrum. Methods Phys. Res. B194 (2002) 177.

9.10 Quantitative Analysis of the Ge Concentration in a SiGe Quantum Well: Comparison of Low Energy RBS and SIMS Measurements

D. Krecar, M. Rosner, M. Draxler, P. Bauer, H. Hutter

Published in: Analytical and Bioanalytical Chemistry, Springer

Volume 384, Number 2, 2006

525 – 530

ORIGINAL PAPER

D. Krecar · M. Rosner · M. Draxler ·
P. Bauer · H. Hutter

Quantitative analysis of the Ge concentration in a SiGe quantum well: comparison of low-energy RBS and SIMS measurements

Received: 1 August 2005 / Revised: 25 October 2005 / Accepted: 25 October 2005 / Published online: 7 December 2005
© Springer-Verlag 2005

Abstract The germanium concentration and the position and thickness of the quantum well in molecular beam epitaxy (MBE)-grown SiGe were quantitatively analyzed via low-energy Rutherford backscattering (RBS) and secondary ion mass spectrometry (SIMS). In these samples, the concentrations of Si and Ge were assumed to be constant, except for the quantum well, where the germanium concentration was lower. The thickness of the analyzed quantum well was about 12 nm and it was situated at a depth of about 60 nm below the surface. A dip showed up in the RBS spectra due to the lower germanium concentration in the quantum well, and this was evaluated. Good depth resolution was required in order to obtain quantitative results, and this was obtained by choosing a primary energy of 500 keV and a tilt angle of 51° with respect to the surface normal. Quantitative information was deduced from the raw data by comparing it with SIMNRA simulated spectra. The SIMS measurements were performed with oxygen primary ions. Given the response function of the SIMS instrument (the SIMS depth profile of the germanium delta (δ) layer), and using the forward convolution (point-to-point convolution) model, it is possible to determine the germanium concentration and the thickness of the analyzed quantum well from the raw SIMS data. The aim of this work was to compare the results obtained via RBS and SIMS and to show their potential for use in the semiconductor and microelectronics industry. The detection of trace elements (here the doping element antimony) that could not be evaluated with RBS in low-energy mode is also demonstrated using SIMS instead.

Keywords Secondary ion mass spectrometry SIMS (68.49.Sf, 82.80.Ms) · Low energy Rutherford backscattering RBS (82.80.Yc) · Quantum well (73.21.Fg) · Ge δ layer (68.18.-g, 68.47.Pe) · Convolution (02.60.Ed)

Introduction

The past few years have shown a period of tremendous growth for the semiconductor materials industry. The development of new materials in the semiconductor industry is one of the fastest growing industrial areas. The main semiconductor materials are silicon, gallium and indium doped with elements like boron, phosphorus and arsenide. Improving the performance of these materials and constructing new materials including heterogeneous structures made from silicon and germanium and its products is one of the main areas of research in the semiconductor industry. Si/SiGe heterostructures (single or multiple quantum wells) are recently developed materials that are not only used to produce microelectronic devices (such as heterojunction bipolar transistors (HBTs), resonant tunneling diodes (RTDs), high electron mobility transistors (HEMT)) [1] but also to produce quantum cascade lasers (near- to far-infrared and semiconductor diode lasers) [2].

Because of the gradually decreasing sizes of microelectronic semiconductor devices, methods that enable concentration depth profiles to be measured with a depth resolution of only a few nanometers need to be developed, for both production and in analytical purposes. However, the correct interpretation of analytical data is also an essential step towards the development of new devices. The Si/SiGe heterostructures (interfaces) are preferentially manufactured via molecular beam epitaxy (MBE) or metal organic chemical vapor deposition (MOCVD). Techniques commonly used to characterize such structures

---

D. Krecar · M. Rosner · H. Hutter (✉)
Institute of Chemical Technologies and Analytics,
TU Vienna, Getreidemarkt 6/164-AC,
1060 Vienna, Austria
e-mail: h.hutter@tuwien.ac.at
Tel.: +43-158801-15120
Fax: +43-158801-15199

M. Draxler · P. Bauer
Institute of Experimental Physics,
Johannes Kepler University Linz,
Altenbergerstrasse 69,
4040 Linz, Austria

Table 1 Summary of analyzed samples

Layer designation	Layer thickness [nm]	Sample A		Sample B		Sample C	
		Si	Ge	Si	Ge	Si	Ge
cap layer	10	100	0	100	0	100	0
SiGe	25	75	25	75	25	75	25
Doping layer	*15*	*75*	*25*	*75*	*25*	*75*	*25*
SiGe	10	75	25	75	25	75	25
Quantum well	**12**	**100**	**0**	**95**	**5**	**90**	**10**
SiGe	500	75	25	75	25	75	25

The nominal concentrations of the analyzed elements at different depths are presented. All values are expressed as atom percentages (at%). Note that the layer written in italic font is the doping layer; the nominal concentration of the doping element antimony was 0.25 at%. The layer written in **bold font** is the quantum well layer

include high-resolution X-ray diffraction (HRXRD), transmission electron microscopy (TEM), photoluminescence (PL), but the most common used methods are secondary ion mass spectrometry (SIMS) and recently Rutherford backscattering (RBS).

Due to its sufficiently good depth resolution (using low-energy RBS) and its ability to provide standard-free quantification, RBS has been widely used to analyze film thicknesses in the nanometer range. SIMS is widely used as a depth profiling technique, although it has some drawbacks caused by sputter processes. It can be used in combination with mathematical fitting procedures for raw data, such as pointto-point forward convolution [3], to determine layer parameters such as thickness, position, and if standards are used, the concentrations of elements as well as structures (quantum wells).

The aim of this work was to compare RBS data with that from SIMS and so to evaluate the performances as well as the drawbacks of these techniques.

Experimental

The SiGe samples were produced by molecular beam epitaxy (MBE), as described in detail elsewhere [4]. The samples consist of different layers as shown in Table 1 and Fig. 1.

The substrate of the Ge δ layer was a Czochralski-grown Si p-type wafer. One atomic layer of Ge was deposited onto the substrate with MBE. The deposition rates were 0.05 nm s^{-1} for silicon and 0.01 nm s^{-1} for germanium. Fifty nanometers of silicon were also deposited onto the Ge layer with MBE.

The experiments were carried out at the −15° beam line of the 700 keV van de Graaff accelerator at the Institute of Experimental Physics at the Johannes Kepler Universität Linz. The experimental set-up is described in detail in [5], so we only briefly present the experimental parameters here. In our set-up, a cooled PIPS detector with ~12 keV energy resolution was used which is located under the beam at 165° to the beam direction. The solid angle of the detector was 2.5 msr. The pressure in the main chamber was kept at 10^{-9} mbar during measurement. The beam had an elliptical shape and was ~2 mm wide and ~4 mm high. Beam currents were on the order of 1–2 nA. The typical acquisition time of a spectrum was about 1,000 s, so that on average only 1×10^{13} projectiles/cm^2 were needed to collect a complete spectrum. Thus, beam-induced damage of the sample is kept very low.

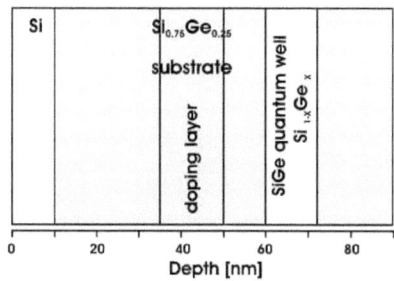

Fig. 1 Schematic layer structure of the samples of Si$_x$Ge$_{1-x}$ analyzed. Sb is added at a low concentration to the doping layer

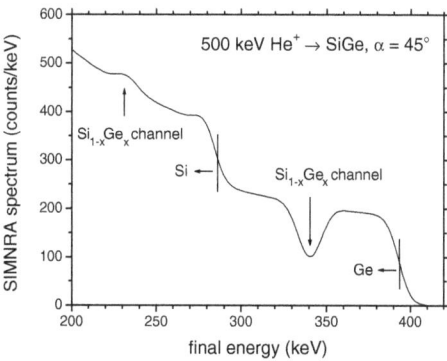

Fig. 2 RBS spectrum for a sample (see also Fig. 1) with x=0.05. This spectrum was obtained by computer simulations using the SIMNRA code for 500 keV He projectiles and for a tilt angle α=45° with respect to the scattering plane

Fig. 3 Measured (*open diamonds*) and SIMNRA simulated (*solid line*) RBS spectra for sample B for 500 keV He ions and an angle of incidence α=45°

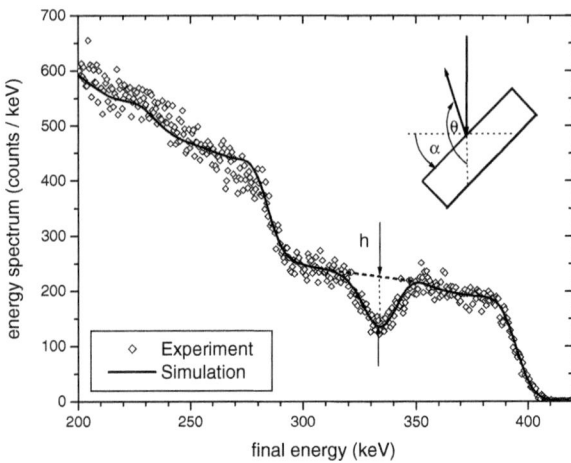

The SIMS instrument used throughout this investigation was an upgraded Cameca IMS 3f. The improvements to the device were mainly associated with the primary section: an additional primary magnet enables the use of a fine-focus Cs$^+$ ion source as well as a duoplasmatron source (in our case generating O$_2^+$ primary ions); a digital scan generator replaced the original beam deflection. Within the scope of this work, O$_2^+$ ions, accelerated to energies of 5.5 keV, resulting in a primary current of 30 nA, and focused to a spot diameter of 50 μm, were used for the production and subsequent detection of positive secondary ions. The primary ion beam was projected onto an area of 350×350 μm^2 with an analyzed area 60 μm in diameter selected via an aperture diaphragm [6]. The recorded masses were ^{16}O$^+$, ^{30}Si$^+$ and ^{70}Ge$^+$ and the response function of the SIMS device (the germanium δ layer) was detected along with the other measurements.

Results and discussion

Rutherford backscattering (RBS) is one of the techniques most frequently used to perform nondestructive, reference-free quantitative analysis of composition, thickness and depth profiles of thin films or interfaces [7, 8]. Typically MeV ions of H$^+$ or He$^+$ are accelerated toward the target and the energies of the backscattered ions are analyzed. The energy of a backscattered ion depends on the mass of the atom that deflects it. The most effective way to calculate RBS spectra over a wide energy range is to use computer simulation programmes like SIMNRA [9]. The latest version of the SIMNRA code considers non-Rutherford cross-sections, isotope effects, realistic stopping powers, energy loss straggling, surface roughness, dual and multiple scattering, and so forth [9]. Since it is representative of all of the investigated samples, the results for sample B will be explained below in detail, but the results for all three samples will be presented at the end. Figure 2 shows a typical simulated spectrum for a sample (as depicted in Fig. 1), with x=0.05, for 500 keV He ions and an angle of incidence of 45°. By choosing a primary energy close to the stopping maximum, the required depth resolution (~10 nm) is obtained for a 45° tilt of the target with respect to the incoming beam. The chosen experimental parameters represent a good compromise for performing quantitative composition analysis in the Si$_{1-x}$Ge$_x$ channel (the quantum well).

Given the optimal measuring parameters, all of the samples presented in Table 1 were analyzed via low-energy RBS. After calibrating the energy of the multichannel analyzer (two spectra of a thick Cu film at different energies, 400 and 500 keV), RBS spectra were obtained for the samples at 500 keV. For this sample (B), the measured RBS spectrum is presented in Fig. 3 together with simulated spectra obtained for optimized depth profile parameters (solid line). As can be seen from Fig. 3, the Si cap layer is thicker than expected, and therefore the measured position of the Ge high-energy edge occurs at a lower

Table 2 Summary of results obtained via RBS for sample B (all values are expressed as percentages)

Nominal values			Measured values		
Layer thickness [nm]	Si	Ge	Layer thickness [nm]	Si	Ge
10	100	0	13	100	0
25	75	25	25	82	18
15	75	25	15	80.5	19.5
10	75	25	10	80	20
12	95	5	12.4	96	4
500	75	25	500	81	19

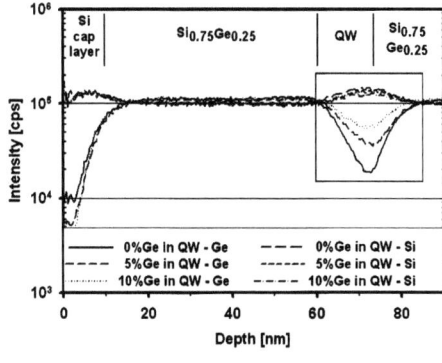

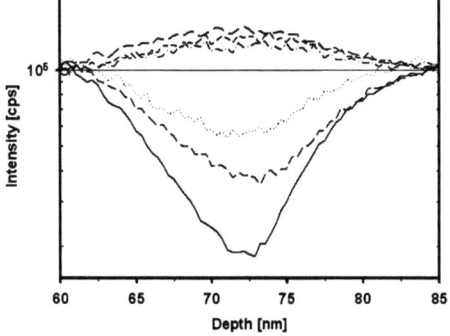

Fig. 4 SIMS depth profiles of the investigated samples. The primary ions used were O_2^+ with an energy of 5.5 keV. The angle of incidence was 42.4°. The primary ion beam was scanned over an area of 350×350 μm² with a primary ion current of 30 nA; secondary ions were collected from the center (∅ 60 μm) of the scanned area. The *upper chart* shows the Si and Ge depth profiles of all of the investigated samples. For better differentiation of the depth profiles, results for the main part investigated, the quantum well, are presented in the *lower chart*

energy than expected. The compositions of the individual layers are optimized such that the plateaux due to scattering from Si and from Ge are both reproduced by the SIMNRA simulation. The content of the doping element Sb was low enough (<0.25 at%) that it should not significantly contribute to the measured spectrum. The results are shown in Table 2 together with the nominal values.

Figure 4 shows in depth SIMS Si and Ge profiles for the investigated samples. The 0 at% Ge concentration in the cap layer (first 10 nm) is indicated, but due to some drawbacks of the SIMS method (atomic mixing, implantation of primary ions), which are especially apparent with the first layers, it is not seen clearly. The decrease in the Ge concentration in the quantum well, as seen in the RBS spectrum (Figs. 2 and 3), is also clearly visible. The channels (quantum wells) in the samples should give sharp interface edges but SIMS measurements of such thin layers incur

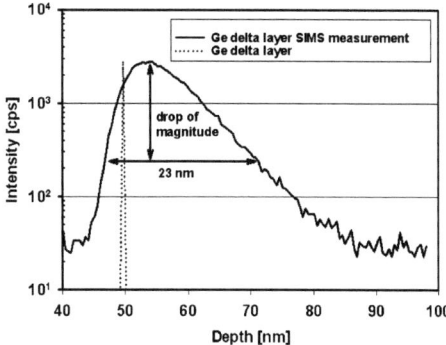

Fig. 5 SIMS depth profile of the Ge δ layer compared with the as-manufactured Ge δ layer

major problems with profile distortion induced by the primary beam. Measurements of such samples are a challenge for SIMS because the layer thickness, of the quantum well here, is less than the depth resolution of the method. Therefore, mathematically, this depth profile distortion can be explained as a convolution of the real nature of the sample with a SIMS response function.

The SIMS depth profile of the Ge δ layer, which is used to calculate the Ge concentration, the position and the thickness of the quantum well is represented in Fig. 5. This Ge monolayer is buried beneath a capping layer of Si with a nominal depth of 50 nm. The Si capping layer is critical to achieving sputter equilibrium until the δ layer is reached during the depth profiling. The Ge δ layer was measured on

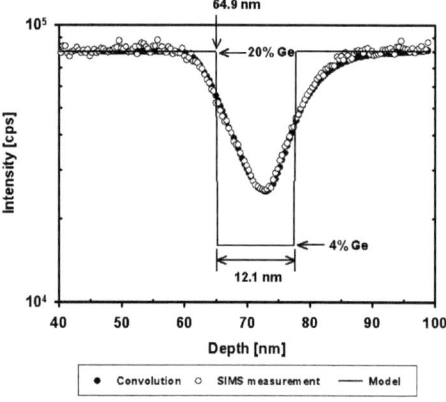

Fig. 6 Demonstration of the convolution procedure for sample B. A model is convolved point-to-point with the SIMS response function, the Ge δ layer, and modified until the best agreement with the measured profile is achieved. The channel width is determined as 12.1 nm (RBS: 12.4 nm)

Table 3 Comparison of SIMS and RBS results

Sample	Nominal depth	Measured depth RBS	Measured depth SIMS	Nominal width	Measured width RBS	Measured width SIMS
A	60.0	62.4	63.9	12.0	11.6	11.4
B	60.0	63.0 ±0.2	64.9 ±0.5	12.0	12.4 ±0.2	12.1 ±0.5
C	60.0	58.0	61.7	12.0	13.0	12.1

All values are expressed in nanometers. The uncertainty is ±0.2 nm for the RBS depth scale and ±0.5 nm for the SIMS depth crater measurements

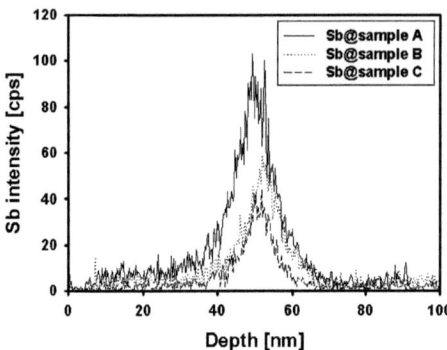

Fig. 7 SIMS depth profiles for the doping element antimony in the investigated samples

the same day and in the same sample holder to ensure that the same conditions were used for all measurements.

Inverse modeling of the distorted SIMS depth profiles was performed by convolving an assumed undistorted profile (the model) with the SIMS response function. For this procedure, the SIMS depth profile of the Ge δ layer (Fig. 5) was normalized and then used as the response function of the SIMS device and subsequently convolved point-to-point with the model profile. A rectangle model profile was used here due to knowledge of the sample production (MBA) [4]. Due to the method of sample production it was assumed that the concentrations of both Si and Ge in front of and behind the quantum well were (nearly) constant. This was subsequently verified by the straight SIMS signals of these of both elements in these regions. The resulting convolution profile was manually fitted to the measured Ge profile by varying the following parameters of the model until the best fits were achieved for both graphs: Ge intensity before, after and in the quantum well, position and width of the quantum well. The result obtained for sample B is displayed in Fig. 6. The Ge channel starts at a depth of 64.9 nm and its width is 12.1 nm. Samples A and C were investigated in the same way. The overlap between the measured and convolved profiles depends somewhat on the model used, but more so on the measured response function. The measurements of the samples must be taken under the same conditions as the measurement of the response function, otherwise no overlap between these two profiles is possible. The accuracy of the results obtained depends upon the precision and the dynamic range of the response function (two orders of magnitude here).

Tables 3 and 4 give a comparison of the nominal Ge concentration values in the quantum well and the concen-

Table 4 Comparison of SIMS and RBS results

Probe	Nominal Ge concentration in the QW	Ge concentration (RBS)	Ge concentration (SIMS)
A	0	0	0
B	5	4±0.5	4±0.5
C	10	9.5	9.5

All values are expressed as percentages. The uncertainties in the RBS and SIMS concentration values are ±0.5%

tration values determined by low-energy RBS and SIMS. The nominal and obtained values for the depth and thickness of the quantum well are also presented. The measured values obtained for the quantum well via SIMS are larger than the RBS results while the measured SIMS widths are smaller than those obtained with RBS, but no significant reasons for these trends were found.

As can be seen in Table 1 and Fig. 1, there is a doping layer within the sample structure. The doping element was antimony (n-emitter). Figure 7 shows SIMS depth profiles of antimony in the investigated samples, which were not obtained using low-energy Rutherford backscattering spectra. The nominal concentration of antimony in the doping layer was 0.25 at%. Figure 7 plots antimony intensity versus depth and not concentration because of the lack of an exact antimony standard. The concentrations calculated using the standard of antimony implanted in cz silicon (note: real matrix $Si_{75}Ge_{25}$) were 0.02 at%. Although, the matrix materials are different, there should not be such a large difference (one order of magnitude) between the nominal and calculated value.

Conclusion

In this paper we have shown that although low-energy RBS and SIMS are very useful standalone techniques, they are far more useful if they are applied together for the quantitative analysis of Ge in samples containing $Si_{1-x}Ge_x$ layers and interfaces (quantum well). The results obtained using these two techniques—the depths, thicknesses and Ge concentrations of the quantum wells—as shown in Tables 3 and 4, agree very well. These results (RBS and SIMS) also agree well with the nominal values for these parameters.

If the energies of the primary ions used for SIMS measurements are decreased, atomic mixing can be reduced and a better depth resolution can be obtained. However, low-energy measurements are only applicable if the analyzed layer is near to the surface and not buried more than 50 nm, as shown in this work. Measurements of the layers in question made at lower energy also have the drawbacks of bad beam geometry and very long measuring times. The longer the measuring time, the more measurement artefacts occur and the higher the costs.

The analytical problems increase as the sizes of the devices produced by the microelectronics industry decrease. It has reached the point that even SIMS using lower primary ion energy is not always able to analyze the layers of interest, so it is necessary to use mathematical models.

In this work we have shown that low-energy RBS provides good depth resolution and enables us to determine the position, width and quantitative composition of the quantum well. Additionally, we have shown that SIMS using higher primary ion energy can also be applied to perform this task. We have seen that atomic mixing, a drawback of using SIMS, which strongly limits the depth resolution and leads to depth profile distortion, can be compensated for during data analysis via mathematical procedures, including point-to-point convolution, as used within this work.

Furthermore, we have shown that SIMS can detect the doping element antimony, which other techniques fail to do, which shows that SIMS remains an indispensable technique in the semiconductor industry and in applied research.

Acknowledgements The authors want to thank M. Mühlberger and F. Schäffler for the production and making the samples available for us.

References

1. Fernández JM, Hart L, Zhang XM, Xie MH, Zhang J, Joyce BA (1996) J Cryst Growth 164:241
2. Maier M, Serries D, Geppert T, Köhler K, Güllich H, Herres N (2003) Appl Surf Sci 203–204:486
3. Dowsett MG, Barlow RD, Allen PN (1994) J Vacuum Sci B 12(1):186
4. Schäffler F, Többen D, Herzog H-J, Abstreiter G, Holländer B (1992) Semicond Sci Technol 7:260
5. Kastner F (1996) PhD thesis. Johannes Kepler Universität Linz, Austria
6. Hutter H (2002) Dynamic secondary ion mass spectrometry. In: Bubert H, Jenett H (eds) Surface and thin film analysis. Wiley-VCH, Weinheim, pp 106–121
7. Chu WK, Mayer J, Nicolet M-A (1978) Backscattering spectrometry. Academic, New York
8. Tesmer JR, Nastasi M (1995) Handbook of modern ion beam materials analysis. Materials Research Society, Pittsburgh, PA
9. Mayer M (2002) Nucl Instrum Meth B 194:177

Die VDM Verlagsservicegesellschaft sucht für wissenschaftliche Verlage abgeschlossene und herausragende

Dissertationen, Habilitationen, Diplomarbeiten, Master Theses, Magisterarbeiten usw.

für die kostenlose Publikation als Fachbuch.

Sie verfügen über eine Arbeit, die hohen inhaltlichen und formalen Ansprüchen genügt, und haben Interesse an einer honorarvergüteten Publikation?

Dann senden Sie bitte erste Informationen über sich und Ihre Arbeit per Email an *info@vdm-vsg.de*.

Sie erhalten kurzfristig unser Feedback!

VDM Verlagsservicegesellschaft mbH
Dudweiler Landstr. 99
D - 66123 Saarbrücken
www.vdm-vsg.de

Telefon +49 681 3720 174
Fax   +49 681 3720 1749

Die VDM Verlagsservicegesellschaft mbH vertritt

Printed by Books on Demand GmbH, Norderstedt / Germany